Transgenics in Dispute

Cristiano Luis Lenzi

Transgenics in Dispute

Political conflicts in the commercial liberation of GMOs in Brazil

 Springer

Cristiano Luis Lenzi
School of Arts, Sciences and Humanities
Universidade de São Paulo
São Paulo, São Paulo, Brazil

This book is a translation of the original Portuguese edition "Transgênicos em disputa: Os conflitos políticos na liberação comercial dos OGMs no Brasil" by Cristiano Luis Lenzi, published in Brazil by Appris Editora in 2020.

ISBN 978-3-031-21791-3   ISBN 978-3-031-21792-0   (eBook)
https://doi.org/10.1007/978-3-031-21792-0

© The Editor(s) (if applicable) and The Author(s), under exclusive license to Springer Nature Switzerland AG 2022

English translation of the 1st original Portuguese edition published by Appris Editora, Curitiba, 2020.
This work is subject to copyright. All rights are solely and exclusively licensed by the Publisher, whether the whole or part of the material is concerned, specifically the rights of reprinting, reuse of illustrations, recitation, broadcasting, reproduction on microfilms or in any other physical way, and transmission or information storage and retrieval, electronic adaptation, computer software, or by similar or dissimilar methodology now known or hereafter developed.
The use of general descriptive names, registered names, trademarks, service marks, etc. in this publication does not imply, even in the absence of a specific statement, that such names are exempt from the relevant protective laws and regulations and therefore free for general use.
The publisher, the authors, and the editors are safe to assume that the advice and information in this book are believed to be true and accurate at the date of publication. Neither the publisher nor the authors or the editors give a warranty, expressed or implied, with respect to the material contained herein or for any errors or omissions that may have been made. The publisher remains neutral with regard to jurisdictional claims in published maps and institutional affiliations.

This Springer imprint is published by the registered company Springer Nature Switzerland AG
The registered company address is: Gewerbestrasse 11, 6330 Cham, Switzerland

To my mother Irma, and Dayane.

Acknowledgments

Many works that address the conflict over RR soybean in Brazil take it from a dualism that separates critics from defenders. Although this confrontation helps to capture what happened in the period when the conflict began, such a view tends to simplify the complexity of the public debate on GM soybean. More than transgenics themselves, the dispute was also spurred by issues of distribution and consumption, and, more generally, by concerns raised by the regulation of technological innovation in the country. For this reason, in this book, the conflict is examined from three distinct axes of analysis. Each one is explored in different chapters of the book. The text is the result of a research developed between 2009 and 2011 and that, during this period, had the financial support of the São Paulo Research Foundation (Fapesp). Thus, I am grateful to this institution for the support given to me at the time. Some of the analyses contained herein were partially published in national academic journals in Brazil. However, unlike them, the present book takes this conflict as a whole and offers a wider range of information and analyses.

Contents

1	**Introduction**...	1
	References..	5
2	**Environmental Policy Process: From Linear to Discursive Model**...	7
	2.1 Analysis of the Environmental Policy Process	7
	2.2 The Language of Environmental Policy	9
	2.3 Discourse, Frames, and Story Lines	12
	2.4 Methodological Considerations...........................	16
	References..	16
3	**Brave New World of Biotechnology**...........................	19
	3.1 The Radical Nature of Biotechnology	22
	3.2 The Risks of Biotechnology..............................	26
	3.3 The Ethics of GM Food	29
	3.4 Agricultural Innovation and Distributives Issues...............	31
	3.5 Public Participation for What?............................	32
	3.6 Regulating GMOs	34
	3.7 The Politicization of Science.............................	36
	3.8 The Labeling Conflict	37
	3.9 The Commercial Release of RR Soybean in Brazil.............	38
	3.10 Final Considerations	42
	References..	43
4	**A Territory Free of Transgenics: The Conflict over the Release of RR Soybean in Southern Brazil**	47
	4.1 The Concept of Environmental Justice	49
	4.2 Distributive Impacts of Agricultural Biotechnology	52
	4.3 GMOs and Environmental Justice in Brazil	52
	4.4 Trade-Off: When Everybody Wins	54
	4.5 GM Seeds and Land Reform	57
	4.6 A Late Threat from the Green Revolution...................	61
	4.7 Fiscalizing (Il)Legal GM Crops...........................	63

	4.8 Back to the Beginning	78
	4.9 GM Soybean: A Friend of Agrarian Reform?	84
	4.10 Final Considerations	89
	References	90
5	**Science in Dispute: Sound Science and the Conflict over Risk Analysis**	93
	5.1 A World Where Everything Becomes Ideology	96
	5.2 When Security Is Not Secure Enough	97
	5.3 Universalism and Equivalence in Risk Analysis	99
	5.4 Epistemological Uncertainty and the Precautionary Discourse	102
	5.5 CTNBio, Regulation and Scientific Uncertainty	104
	5.6 Does the Absence of Evidence Indicate the Absence of Risk?	106
	5.7 Beyond Risk Analysis?	109
	5.8 Final Considerations	114
	References	115
6	**Labeling as Precaution: Substantial Equivalence and the Conflict over Labeling**	119
	6.1 Labeling Risks	122
	6.2 Defending the Consumer's Right in the Absence of Danger	124
	6.3 Consumer Choice and Environmental Risk	128
	6.4 The Ideological Conflict About Labeling	131
	6.5 The Policy of Substantial Equivalence	133
	6.6 Defending the Consumer from Himself	136
	6.7 An Environmental Utopia	138
	6.8 Final Considerations	139
	References	140
7	**Regulation Made in the United States: Regulatory Polarization and the Brazilian Case**	143
	7.1 Regulatory Polarization	143
	7.2 The European Case	144
	7.3 The American Case	149
	7.4 Brazil, Europe, and United States	151
	7.5 Final Considerations	160
	References	161
8	**Conclusion**	163
	References	166
Index		167

About the Author

Cristiano Luis Lenzi is a sociologist and Professor of Environmental Sociology in the School of Arts, Sciences, and Humanities at the University of São Paulo.

Chapter 1
Introduction

In the late 1990s, the commercial release of Roundup Ready (RR) soybean triggered a conflict involving environmental NGOs, small farmers' organizations, farmers, and the different levels of government in the state of Rio Grande do Sul. On one side of the conflict were farmers favorable to the use and commercialization of RR soybean in agriculture and, on the other, civil society organizations, political leaders, and the state government itself that sought to bar the commercialization of these products in the southern region of the Brazil. After this initiative, similar movements also took place in other parts of the country, which stimulated, from 2001 onwards, the emergence of restrictions for GM crops in other Brazilian states. The apex of the conflict occurred in 1999, when the Olívio Dutra (PT) government in RS launched the campaign with the slogan "Um Território Livre de Transgênicos." From a political perspective, one can find significant political effects produced by the campaign. With it, the south region of Brazil became a symbol of resistance to GMOs and helped broaden the national and global debate on the issue. At the same time, the objective results produced by it were insignificant, if one considers that GM crops have since become an agricultural reality in Brazil.

GMOs are not the only controversy associated with the environmental problems produced by modern society. The controversies that surround them stem from the very complexity that can be associated with them. While many of these problems can be known from specific aspects of the natural world (air, soil, water, etc.), they are not only related in complex ways but are integrated into our very way of life. Environmental problems constitute hybrid phenomena because they involve social relations and nonhuman natural phenomena (Giddens & Sutton, 2012, p. 126), which gives them a higher level of complexity that can be defined here as the number and variety of elements in a system (Dryzek, 2005, p. 08). It is for these reasons that sociologist Ulrich Beck will state that "nature can no longer be understood outside of society, or society outside of nature" (Beck, 1992, p. 80), which, according to him, makes the "destruction of nature" not only the destruction of something that is "out there" but is also a process that is part of the "social, economic, and political dynamics of the human world".

Because of these characteristics, it is not uncommon for environmental problems to come to us through scientific, ethical, and political controversies of different kinds. Governments, civil society, and economic actors clash to determine the lines that define them. They all seek to persuade the public to impose a definition of the problem that is the most attractive to them. As such, these problems cannot be dealt with through a narrow separation between facts and values, because the former are integrated into the interests of the actors moving in conflict. A search for objective facts could not solve such disputes, since what counts as evidence, and how it is to be interpreted, is given by the very interpretative structures in dispute. These disputes are thus influenced by discourses that seek to establish the lines along which environmental problems are defined.

Discourse can be seen, then, as a form of argumentation in which the different participants in political life present justifications for their positions in an attempt to influence the political process in some way. This emphasis on discourse has given rise to what Fischer and Forester (1993) call the discursive turn in political analysis. From this perspective, politics is seen as a process of interpretive struggle to define public issues. A struggle to define the boundaries of the public issues under debate and the imposition of a dominant interpretation for these issues. In this framework, the analysis of what in English is called *frame* becomes an important element for the study of the political process. As Fischer (2003, p. 143) indicates, the most important question for interpretive political analysis is: "how is the political issue being conceptualized or structured by the parties involved in the debate?" The political process can be seen as an essentially discursive process in which existing interpretive structures, defined in this study as frames, provide the symbolic referents for the construction of arguments in political conflicts. Frames can be seen as symbolic constructs that include beliefs, images, and symbols shared by political actors who make use of them in an attempt to understand issues in dispute. They are the frames that make the world meaningful and structure the way one interprets the events of reality around us.[1]

Let us take the example of poverty offered by Rein and Schön (1993, 1994) to further illustrate the point. In public debates about this social problem, there is not just one problem being discussed, but several. In cases involving disputes over public policies for poverty, each party sees the phenomenon in a different way, they claim, making the issue different for the parties in conflict. It should not be said that "we are comparing different perspectives on 'the same problem', because the problem itself has changed" (Rein & Schön, 1993, p. 153). Taking this understanding in

---

[1] In media studies, frame is used interchangeably with the term framing. The term is also sometimes used as a synonym for discourse. However, as will be shown in the first chapter, there are differences between these concepts that must be preserved, although they are integrated with each other. At the same time, the meaning that is given to the term in media studies does not always do justice to the use of the term in political process research. Therefore, this study affiliates itself with the understanding of frames as found in public policy studies.

perspective, the authors then ask what would be the resolution of such disputes when the very frames of the dispute demarcate what should be considered as evidence and how it should be interpreted (Rein & Schön, 1993, p. 145). The arguments applied to the example offered by these authors, in turn, can be applied to the case of commercial release of RR soybean in Brazil. In this last case, there is not just a problem associated with GMOs being examined, but several that arise from the very interpretative structures (frames) that are placed in confrontation.

In this book, the conflict produced by the commercial release of RR soybean is examined from the precepts of discourse analysis of policy. In particular, the conflicts that arose from this decision at the end of the 1990s and its development in the following years are examined. The study focuses on the strategies used by the actors in conflict to legitimize their positions on the issue. In this way, the paper offers a discursive approach to these political disputes that surrounded the commercial release of RR soybean and argues that, in analyzing the dispute, it would be possible to find three different axes of dispute in the conflict. In each, two dominant and distinct groups of actors are examined, which will be referred to in this paper as discursive alliances. The aim of the study is to show, then, that the conflict over commercial release of RR soybean represents a political controversy, in which distinct interpretative structures (*frames*) shape different political discourses about the impacts brought about by the release of GMOs in the country.

The next chapter of the book offers a clarification of some of the concepts that guide the analysis in this study. In this first part, the interpretive turn in political analysis is examined and also its relationship to the concepts of framing, discourse, and story line. The book is thus affiliated to studies that seek to understand political conflicts from a discursive perspective. If, as Fischer (2003, p. 143) indicates, the main question of this type of analysis is "how is the policy issue being conceptualized or 'framed' by the parties to the debate?", in the present study, this same question is changed to: how have the environmental risks associated with GMOs been perceived by the parties involved in the process of commercial release of RR soybean in Brazil? In attempting to answer this question, the study is divided into three distinct analyses that are developed in the Chaps. 4, 5, and 6 of the book.

The third chapter provides an introduction to understanding the controversies associated with biotechnology. In this part of the chapter, some of the reasons why controversies about biotechnology have arisen are analyzed. To do this, some of the reasons that lead to disputes over issues of risk, ethics, fairness, labeling, and regulation in the GMO debate will be examined. The chapter concludes by offering a brief introduction to the conflict that occurred with the commercial release of RR soybean in Brazil at the end of the 1990s. This last part will serve as an introduction to the analyses that will be presented in subsequent parts of the book.

In the fourth chapter, which opens our analysis of the controversy surrounding the commercial release of RR soybean in Brazil, the conflict is examined through a distributional bias. This axis of the conflict will be referred to as the environmental justice (EJ) story line. The position in favor of the commercial release of RR soybean will be referred to in our analysis as the discourse or political alliance of agricultural modernization. The group of actors who sought to paralyze this release will be referred to as the EJ political alliance and discourse. In the country, analyses about GMOs usually focus on aspects involving environmental risks and their labeling process. This is what, in a sense, will be done in the following chapters of the book. But when the interpretation of the conflict is reduced to environmental risks, resistance to GMOs is interpreted almost exclusively as an environmental safety issue. However, the conflict over GMOs in Brazil did not start with these issues at the forefront, as will be shown in this first part. More than issues related to the environmental impacts of RR soybean, the conflict was associated with distributional issues related to agrarian reform. The conflict in the south of the country expresses a cultural peculiarity, since the factors that caused this conflict to erupt in the region are associated with the historical and permanent tensions that occur in the region. This part of the work will thus allow us to understand how the commercial release of RR soybean ended up integrating themselves into an agricultural conflict that has been ongoing in the region for many years. With this, the present chapter brings a contribution to understanding how technological innovations in agriculture can engender already existing conflicts that remain latent in the country.

In the fifth chapter, what will be referred to as the uncertainty story line will be examined. This conflict axis expresses issues of scientific uncertainty associated with the environmental risk of RR soybean. This axis of analysis is formed by the conflict between two political alliances that clashed in this process. The political alliance and discourse of liberation are characterized by a minimal preventive approach. It is represented especially by CTNBio, the Ministry of Science and Technology (MS&T), and other sectors of the scientific, political, and economic fields. In this discourse, the preventive measures that are seen as necessary for the commercial release of RR soybean tend to be reduced to the application of an RA. However, this discursive alliance presents a series of assumptions regarding PP itself and issues involving the perception of risk, science, and uncertainty that distinguish it, in turn, from what will be called the precautionary political discourse or alliance. The precautionary alliance is represented by organizations such as Idec, Greenpeace, and other political actors that offered a differentiated discourse on the application of PP to GMOs. It is characterized precisely by a contestation to risk analysis (RA) in the process of commercial release of RR soybean.

In the sixth chapter, the conflict surrounding the labeling of GM foods will be examined. In the last decade, an increasing number of countries have implemented labeling policies for GM foods. And beyond this measure, the conflicts and tensions that such a policy usually engenders have also become common in these countries. For companies, the label is a central element of product marketing policy and thus tends to be seen as directly influencing consumer decisions. For environmentalists, on the other hand, the label tends to be seen as a space of symbolic struggle and as a means of achieving better regulation of these products. For these reasons, it seems natural that labeling constitutes an important space for struggles that seek to define the commercialization of GMOs. The last chapter of the book examines the unfolding of this conflict in the country. The conflict will be examined from what will be referred to as the labeling story line. The conflict is structured by the confrontation of the political alliance and discourse of conventional labeling, which is, in turn, contested by the political alliance and discourse of precautionary labeling. This conflict represents the axis through which the various structuring issues of the labeling conflict presented themselves in the country. Among the points that will be examined in this story line are the views on the relationship between labeling, precaution, consumer choice, and other points of disagreement. As in the previous chapters, it is possible to verify the existence of two distinct political alliances that are organized around the issue of labeling and that offer different political discourses on the issue. These political alliances and discourses present different premises about labeling and its relation to issues involving the principle of substantial equivalence, science, risk, and nutritional safety. This chapter, therefore, addresses the issue of RR soybean in Brazil as a political controversy in which distinct political actors and their respective worldviews come into conflict, structuring political discourses on GMOs in the public space and thus influencing the decision-making process in Brazilian environmental policy in some way.

In the seventh chapter, a brief comparative analysis of existing regulatory policies in Europe, the United States, and Brazil is presented. With this final analysis, one seeks to situate the Brazilian regulatory policy taking into perspective the regulatory alternatives that exist today in the world. Considering that the United States and Europe are seen today as representing different models in the regulation of GMOs, the proximity of Brazilian regulatory policy to these two alternatives will be examined. At the same time, taking up the analysis developed throughout the book, arguments are offered to help understand why the Brazilian case tends to be closer to one model than the other. In the eighth and last chapter, some of the conclusions drawn from our analysis are presented. Within the limits of what has been examined in this book, the main points of the analysis carried out throughout the book are briefly resumed in order to extract some overview of the GMO regulation process in Brazil.

References

Beck, U. (1992). *Risk society: Towards a new modernity*. Sage.
Dryzek, J. S. (2005). *The politics of the earth: Environmental discourses*. Oxford University Press.
Fischer, F. (2003). *Reframing public policy: Discursive politics and deliberative practices*. Oxford University Press.
Fischer, F., & Forester, J. (1993). Introduction. In F. Fischer & J. Forester (Eds.), *The argumentative turn in policy analysis and planning*. Duke University Press.
Giddens, A., & Sutton, P. (2012). Meio ambiente. In A. Giddens (Ed.), *Sociologia*. Penso.
Rein, M., & Schön, D. (1993). Reframing policy discourse. In F. Fischer & J. Forester (Eds.), *The argumentative turn in policy analysis and planning*. Duke University Press.
Rein, M., & Schön, D. (1994). *Frame reflection. Toward the resolution of intractable policy controversies*. Basic Book.

Chapter 2
Environmental Policy Process: From Linear to Discursive Model

Policy process, implementation, and regulation are some of the terms used to understand how governments seek to respond to the risks generated by techno-scientific development. These different terminologies are used to understand the political process of environmental decision-making and one can consider three dominant approaches of analysis in the literature on the subject. The first one sees the political process as a reflection of social and economic interests, while the second one seeks to evaluate this process from a political community perspective. In the third, politics is seen as a symbolic struggle in which the concept of discourse becomes the central element of the analysis. The present study is oriented on the third and second approaches. Next, a brief assessment of the premises of the discursive approach to environmental politics will be made and then some of the concepts that structure the research are examined. In Chaps. 4, 5, and 6, this theoretical framework will guide the analysis of the conflict involving the commercial release of RR soybean in the country.

2.1 Analysis of the Environmental Policy Process

An influential model of political analysis has been what some authors call the linear model of the political process. According to this view, the policy process follows a rational and predictable development that unfolds in different stages that follow one another. Among the main stages is the creation of a public agenda, the decision-making process, and, finally, implementation. Policy-making is usually seen as being confined to a select group of people in key positions in the government structure. In doing so, this approach implies a top-down reading to understand the decision-making process. The rationality of the political system is reflected in the strategic action of policymakers at the highest level of the process. Policy comes to be seen as a type of problem-solving in which the political process involves identifying and defining policy problems and then establishing decisions and implementing selected measures. Decision and implementation are seen as distinct moments in the process.[1]

Several criticisms have been directed at this type of political analysis. Sabatier and Jenkins-Smith (1993a, b) criticizes this model for not offering an understanding of the forces or mechanisms that drive the political process from one stage to another. Lacking an understanding of these causal factors that lead to this change of stages, this model would lack a clear basis for generating empirical hypotheses.[2] Moreover, many political analysts seek to indicate that the political process is far from following a rigid order of stages. Its development would represent a more complex process in which the different stages would influence each other in a continuous manner in which a series of deviations would prove to be a rule rather than an exception.

The most consensual criticism that one can direct to this type of analysis is present in a distinct approach to the political process that sees the latter as the result of a continuous process of negotiation and bargaining between distinct political actors. This type of analysis is particularly represented by Charles Lindblon, who defined the decision-making process as "science of muddling through" (Lindblon, 1959 apud Keeley & Scoones, 1999). Unlike the linear analytical model of the political process, this approach sees the political process in a bottom-up manner in which the actors involved can influence it in a permanent manner. Today, many analysts are moving from viewing the policy process from a rigid linearity to this viewpoint. The fact that the range of actors involved in the political process is broader than is usually assumed in the stages' model gives the political process the image of a circular rather than linear process. As Reiner and Rabinov (1995, p. 322 apud Juma & Clark, 1995, p. 125) wrote: "The process is not one of a graceful one-dimensional transition from legislation, to guidelines, and then to auditing and evaluation. It is ... circular or looping." Policy, then, does not emerge from a singular point. Before seeing it as the linear unfolding of decision-making stages, it is possible to see it as the broad course of action or a web of interrelated decisions that evolve over time during the process of implementation (Hill, 1997, p. 07). The political process, therefore, rather than being the result of an instrumental execution of rational decisions of policymakers, is itself a political phenomenon in which the process itself, and not only the problems that are being filtered by it, becomes a focus of dispute and conflict of social actors.

Political studies have been characterized by a more heterogeneous framework for conducting political analysis. Among the most striking approaches found in the study of the political process are those that make political language itself an

---

[1] On this approach, see Keeley and Scoones (1995), Sutton (2006), Fischer (2003), and Hill and Ham (1984).
[2] It is important to realize that Sabatier's and Jenkins-Smith (1993a, b) criticism of the stage model conceals a longing for the formulation of a general framework of causal explanations of the policy process. However, this stance has itself been criticized by the post-empiricist analysis of public policy. On this point, see Fischer (2003).

important object of analysis.[3] In sociology and political science, the idea that politics can be seen as a debate between different social actors has been receiving more attention. As MacRae (1993) indicates, political discourse can be seen as a form of argumentation in which different participants in the political game present claims and justifications for their positions.[4] The importance of language in policy analysis has led to the emergence of what Fischer and Forester (1993) call the "argumentative turn in policy analysis." In this view, political decisions are seen as an ongoing process of discursive struggle for the definition of social and political problems. A struggle for their boundaries and the quest for imposing a hegemonic subjective interpretation of the problems under debate. This struggle influences not only the definition, but also the analysis of the problems, the strategies of the actors, and the public's own understanding of what is at stake in the political debate.

A central element of political discourse analysis is its concern with the social meaning that political concepts themselves may have for participants in the political arena. In this framework, the analysis of The English is called frame becomes crucial, as Fischer (2003, p. 143) indicates: "how is the policy issue being conceptualized or 'framed' by the parties to the debate? How is the issue selected, organized, and interpreted to make sense of complex reality?" As the author indicates, in the political process, controversial issues go through a process of selection, organization, and interpretation. Throughout this process of interpretive framing, frames provide the symbolic referents for "analyzing, knowing, arguing, and acting" (Fischer, 2003, p. 143). Next, some of the concepts that will prove crucial in our intended analysis of the political process involving GM seeds in Brazil are examined. Among these are the concepts of discourse, frame, framing, and story line.

2.2 The Language of Environmental Policy

The concept of frame has been used as a means of understanding the way in which public issues are structured in the political arena. Its origin dates back to the work of Erving Goffman, who defined frame as a "principle of organization which govern events – at least social ones – and our subjective involvement in them" (Goffmann, 1974, p. 10). The concept has been used in various fields of knowledge, such as psychology, sociology, political analysis, and conflict studies. Although some definitions of the concept may differ in some points, most of them tend to indicate the

---

[3] In his text *The Policy Process: An Overview*, Sutton (2006) identifies at least four disciplines that currently contribute to the study of the policy process. Among them are (a) political science and sociology, (b) anthropology, (c) international relations, and (d) management.

[4] Or as Majone (1989, p. 1) reminds us: "As politicians know only too well but social scientists too often forget, public policy is made of language. Whether in written or oral form, argument is central in all stages of the policy process. (...) Political parties, the electorate, the legislature, the executive, the courts, the media, interest groups, and independent experts all engage in a continuous process of debate and reciprocal persuasion."

same process: a set of assumptions that individuals have to interpret what happens around them, using language itself as a device of political dispute. Frame represents a set of collective beliefs and, therefore, never a strictly individual phenomenon.

Frames tend to be defined, thus, as "cognitive structures held in memory and used to guide in the interpretation of new experiences" (Gray, 1997), or "structures that we use to name a situation," in order to "name a situation in which we find ourselves, to identify and interpret specific aspects that seem key to us in understanding the situation, and to communicate that interpretation to others" (Buechler apud Kaufman et al., 2003).[5] In the context of social interactions, such interpretative schemas allow us to answer the question "what is happening here?" (Goffmann, 1974, p. 25). Frames are symbolic constructs that include beliefs, images, and symbols shared by a particular group of social actors. It is these frames that make the world meaningful and structure the way we interpret the events of reality that emerge to us. An important aspect of the concept of frame is that it leads our interpretation from a predefined pattern. In this sense, frame is associated with the phenomenon of individual and collective memory. According to Triandafyllidou and Fotiu (1998): "people tend to order experience by relating it to an already known pattern. Perceptive elements are recognised by reference to a pre-existing cognitive structure."

It is important here to make a distinction between frame and framing, since these terms have been used interchangeably in the literature. While several authors use these concepts as synonyms, other works distinguish between them. While a frame does not have a direct interactive consequence, Kaufman and Smith (1999, p. 167) argue that framing" is "the act of deliberately crafting a frame for oneself, or more often, for the benefit of an audience such as a counterpart in negotiations, a constituency, or "the public."" Different social actors, including interest groups and politicians, engage in "careful framing" in order to persuade their interlocutors about issues in dispute (Kaufman & Smith, 1999, p. 167). But this way of seeing the concept, although it seems useful, offers to the term an overly narrow meaning, since it restricts its use to a purely strategic behavior. Framing would imply a situation in which human beings are aware of the assumptions that govern their discourse and use these same assumptions in a deliberate way to influence other actors in the political game. However, as will be indicated below, this deliberate and strategic use is in many cases nonexistent. Although many works emphasize this aspect, this should not be considered as a necessary condition of the interpretative structuring process in the political process. According to Entman (1993, p. 52), those who communicate produce judgments, conscious and unconscious, when deciding what to say to the audience. In these circumstances these actors are guided by the frames that organize existing belief systems.

In this case, if frame can be seen as a set of beliefs that structure our interpretation of reality, framing can be seen as the very process by which this interpretation is developed, whether consciously or unconsciously, without us having to print to

---

[5] For an exposition of these and other concepts in the literature.

2.2 The Language of Environmental Policy

the subject a strategic dimension. Thus, framing is "a way of selecting, organizing, interpreting, and making sense of a complex reality to provide guideposts for knowing, analyzing, persuading, and acting" (Rein & Schon, 1993, p. 146).

As indicated earlier, frame can be seen as an interpretive schema that remains latent in individuals and social groups, while framing is the process of constructing interpretation. Frame can be seen as the set of assumptions that organize our interpretation, while framing is the way these assumptions are activated in political debate. Framing can be seen as the process by which the beliefs associated with the frame are therefore activated. The concept indicates the process of interpretation in action that involves a process of selection and salience. As Rein and Schon (1994) note, the task of making sense of a complex reality requires an operation of selectivity and organization, which is, in this case, what "framing" means.

Framing involves two distinct processes: naming and selection. When a name is assigned to a problematic political issue, such process induces us to focus our attention to certain aspects, causing other aspects to be neglected:

> The name assigned to a problematic terrain focuses attention on certain elements and leads to neglect of others. The organizing of the things named brings them together into a composite whole. The complementary process of naming and framing socially constructs the situation, defines what is problematic about it, and suggests what courses of action are appropriate to it. It provides conceptual coherence, a direction for action, a basis for persuasion, and a framework for the collection and analysis of data – order, action, rhetoric, and analysis. (Rein & Schon, 1993, p. 153)

Framing involves giving more prominence and salience to certain aspects of an issue than others. Salience is defined here as the act of making "a piece of information more noticeable, meaningful, or memorable to audiences" (Entman, 1993, p. 53). In this case, framing serves as a means of directing our attention and emphasizing objects, situations, or events of reality in a selective way. Functioning as an interpretative scheme, it allows us to create a general diagnosis that goes from defining the problem to prescribing solutions (see Table 2.1).

The framing process finds a parallel in Hajer's (1995) concept of story line. In *The Politics of Environmental Discourses*, Hajer indicates that a story line "is a generative sort of narrative that allows actors to draw upon various discursive categories to give meaning to specific physical or social phenomena" (1995, 56). Story line indicates a series of ideas or ways of seeing associated with a discourse. On the other hand, lines-story also indicate that "different discursive elements are presented as a narrative, or story line, in which elements of the various discourses are combined into a more or less coherent whole and the discursive complexity is concealed" (Hajer, 1993, p. 47). As a type of narrative, story lines allow the actors in a conflict to be positioned through the attribution of "blame" or "responsibility" and

Table 2.1 Dimensions of the framing process

Diagnosis	Responsibility	Prescription
What is happening?	Who is responsible?	What should be done? Who should do it?

Source: Author

where it becomes possible to impute to the issues in dispute characteristics such as urgency or seriousness (Hajer, 1995, p. 65).

As an example, one can take the case of rainforest destruction offered by the author. For a systems ecologist, the value associated with the forest integrates into the mathematical equations, as it will also present itself as a component for their integrated ecosystem view. But organizations like World Wildlife Fund may be more concerned with the moral problem of forest destruction. Other actors, on the other hand, may present themselves as being concerned with the connection between forest life and the culture of indigenous peoples. In these cases, as Hajer (1995) points out, each actor has a different cognitive and social point of view, but they all contribute to the production, even if in a particular way, of the story line of tropical forest destruction. Together, all these actors are part of the same discursive alliance and tend to contribute to what can be called the story line of forest destruction. And this occurs even though each of them does so in a different way and offers a particular discourse for it. If the interpretation of what Hajer (1993) has to say is correct in this analysis, we can say that the different ideas, arguments, and discourses that coalesce in a given political conflict (coalesce to the extent that they offer a support for a particular point of view in addressing a political problem) shape a line-story. This seems to indicate two distinct ways of capturing the idea of story line. In the first, story line tory line are ways of narrating events, which belong to the same discourse. In the second case, story line indicates a situation in which "elements of several discourses are combined" (Hajer, 1993, p. 47).

A discourse coalition, a definition that will also be taken here to guide the analysis of the Brazilian conflict over GMOs, is, in turn, "the ensemble of a set of story lines, the actors that utter these story lines, and the practices that conform to these story lines, all organized around a discourse" (Hajer, 1993, p. 47). A discourse coalition suggests that the construction of a particular policy issue is created and sustained by political actors who "form specific coalitions around specific story lines" (Hajer, 1993, p. 47). Story lines are the means by which, according to Hajer (1993), actors seek to impose a view of reality on other political actors.

2.3 Discourse, Frames, and Story Lines

Discourse and frame seem to be similar in many ways. Both are made up of ideas, beliefs, rules, symbols, and metaphors. Both make it possible to give meaning to events and phenomena in the world and both make it possible to interpret events and situations along interpretive lines. Therefore, it is important to make a brief comparison between these concepts to examine some of their differences. Let us take, for this, the concept of discourse and compare it with the concept of frame. As defined by Hajer: "Discourse is here defined as an ensemble of ideas, concepts, and categories through which meaning is given to phenomena." For Hajer (1993, pp. 45–46), discourses structure the way we interpret social and political problems.

2.3 Discourse, Frames, and Story Lines

In the process, they induce our perception of reality to emphasize certain things (salience) to the detriment of others.

The following passage indicates the similarity that Hajer and Versteeg (2005) establishes between discourse and frame. For Hajer and Versteeg, there are many studies that seek to show us how actors in conflict seek to exercise power by trying to impose a "*particular frame or discourse* onto a discussion" (emphasis added) (Hajer & Versteeg, 2005, p. 177).

Therefore, the similarity between the concept of frame and discourse is evident. For Rein and Schon (1993, p. 145), in the frame, "facts, values, theories, and interest are integrated." To Hajer the discourse, in turn, can present "normative or analytic convictions" (1993, p. 45). For Rein and Schon (1993), the frame establishes what counts as evidence and how it should be interpreted. For Hajer (1993, p. 46), the discourse "forms the context in which phenomena are understood and thus predetermines the definition of the problem." Therefore, the discourse structures the problems and "provides the tools with which problems are constructed" (Hajer, 1995, p. 46). For Rein and Schon (1993, p. 153), the frame, through a process of naming and structuring, allows us to construct a situation by establishing what is problematic in it. The similarities between these concepts pose us, then, a challenge. After all, should discourse and frame be seen as interchangeable concepts?

The difference between these concepts is not always easy to be perceived, which favors the vagueness in the use of both. The concepts of frame and framing are generally used interchangeably with the concept of discourse. Bleich (2002), for example, notes a similarity between the concepts of belief system, political paradigm, frame, and referent. Authors such as Weale (1992) use the terms ideology and belief systems as equivalent terms. Purvis and Hunt (1993) also point out the similarity between discourse and ideology. Looking at this whole framework and making a summation of these approximations, it is possible to say that the concepts of belief system, political paradigm, ideology, discourse, frame, and referent are used for very similar purposes in research on the political process. In some, at least some of them, they are used interchangeably, even though it is possible to find specificities in the way each one of them is employed in studies about the political process. The difference in some of these terms used in social research seems to reside in small analytical details. However, an effort will be made to understand the difference between discourse, frame, and framing in the following. This will help us understand why these concepts are integrated with each other, but at the same time have differences that must be considered.

The passages above are more than enough for us to see how these concepts can be seen as similar. But there is a crucial difference that can be established between them. An important feature of the frame is that it presents itself as a cognitive structure preexisting the circumstances of social interaction, as indicated earlier. It is because a frame shows itself as a reference structure accessible to the individual before the process of social interaction takes place that it becomes an important reference point for the person. This "pre-given" character is fundamental for us to understand the way in which frames exert their influence in the social world and social relationships. The fact that this preestablished structure exists allows different events to be filtered through the familiarity that can be created between what is already known and the new event that is yet to be interpreted. As Triandagyllidoi and Fotiou (1998, p. 2) indicate: "people tend to perceive selectively since they are attracted by those perceptive elements that are more familiar to them." In this way, the "elements that fit into *pre-existing cognitive frames* are more easily recognisable" (italics added). This view is also expressed by Rein and Schön (1994, p. 34) when they state that:

> The frames that shape policies are usually tacit, which means that we tend to argue *from* our tacit frames *to* our explicit policy positions. Although frames exert a powerful influence on what we see and how we interpret what we see, they belong to the taken-for-granted world of policy making, and we are usually unaware of their role in organizing our actions, thoughts, and perceptions.

This element exists in the conception of frame that generally disappears in the concept of discourse, because the latter refers only to the words that were said or written texts (Wood & Kroger, 2000, p. 55). As preexisting structures of interpretation, the frames are key elements in structuring interactive patterns and, as such, indicate that certain meanings that structure a social interaction are previously established. Giddens (1984), for example, when using the referential of Goffman's theory in his structuration theory, tells us that the rules we apply reflexively in copresence contexts never have implications for specific encounters, but operate in a way that reproduces the existing patterns of encounters that people perform across time and space (Giddens, 1984, p. 89). Furthermore, although a frame may present itself in the realm of discourse, its existence takes place much more in the realm of social practice. According to him:

> Whenever individuals come together in a specific context they confront (but, in the vast majority of circumstances, answer without any difficulty whatsoever) the question 'What is going on here?' 'What is going on?' is unlikely to admit of a simple answer because in all social situations there may be many things 'goin on' simultaneously. But participants in interaction address this question characteristically on the level of practice, gearing their conduct to that of others. Or, if they pose such an question discursively, it is in relation to one particular aspect of the situation that appears puzzling or disturbing. (Giddens, 1984, p. 87)

Therefore, the constitutive rules of interpretive schemes (frames) are tacitly followed and not formulated in terms of discourse (Giddens, 1984, p. 89). This "natural" character of frames is often found between the lines of what is said and written. They constitute what is "taken for granted." These are premises that are not necessarily made explicit in what is said or spoken. It is these natural elements that make up the frame of environmental discourses. These natural elements make up a set of basic beliefs without which a discourse cannot come into existence. Rein and Schon (1994) signal something similar, since, as seen above, they indicate that frames belong to the taken-for-granted world of policy-making. Which means, as they point out, that we are generally not aware of how frames operate to organize our actions, thoughts, and perceptions (Rein & Schon, 1994, p. 34). In turn, they note that in order to become aware of the different viewpoints that exist in a political

2.3 Discourse, Frames, and Story Lines

controversy, it is necessary to become aware of our frames, which in turn requires that we must "construct them, either from the texts of debates and speeches or from the decisions, laws, regulations, and routines that make up policy practice" (Rein & Schon, 1994, p. 34).

Discourses can take shape thanks to the existence of these basic elements and can only construct their narratives by virtue of their existence. In this study, a frame can be equated, then, with what in Dryzek's (2005) approach to discourse is called a discourse ontology. Thus, a frame can be considered to constitute the basic entities of a discourse, and the latter can be seen as the operationalization of these basic entities in the process of social interaction. Discourse emerges through the process of conversation and will take concrete form through words, concepts, arguments, written texts, etc. The classification that Rein and Schon (1993, p. 145) make becomes pertinent in this study. They define political discourse as the "interactions of individuals, interest groups, social movements, and institutions through which problematic situations are converted to policy problems, agendas are set, decisions are made, and actions are taken".[6] The frames shape the constructions of the political problems from this process of interaction. The conception of discourse that is incorporated in this work is based on this interactionist model. As Rein and Schon (1993) indicate, interaction is the origin and basis of discourse.

Frame refers to the set of beliefs and assumptions that govern the process of interpretation of social actors. This system does not necessarily have its effect exclusively through communication, but can also be expressed through the practices. Frame here represents a cognitive and interpretive schema that guides existing social interactions. Discourse is not only a means by which this frame can be evaluated, but also serves as a means by which this frame is produced and reproduced by language in the process of social interaction. In the former case, discourse comes to function as a methodological device (for the researcher) and, in the latter, as a means and result of the reproduction of the interpretative schemes (frames) that human beings use in their social interactions. Discourse emerges from the dialogue and interactions of social and political actors in social life. For this study, discourse and frames are, then, elements that integrate each other. Frames are related to tacit assumptions used by social actors, while discourse is the transformation of these assumptions through language itself in the process of social interaction. So, for this work, discourse emerges when frames are constructed through "texts of debates and speeches or from the decisions, laws, regulations, and routines that make up policy practice" (Rein & Schon, 1994, p. 34).

---

[6]One difference that seems to mark the work of Rein and Schon (1993) from authors like Hajer (1993) is that the former tends to see discourse itself as a type of social interaction process. This is not very evident in Hajer's (1993) work since he seems to reduce discourse only to the elements that arise from the interaction of social actors (ideals, arguments, metaphors, etc.).

2.4 Methodological Considerations

The analysis presented in Chaps. 4, 5, and 6 draws on various sources of information. The first and main one is the work already carried out that seeks to examine the conflict over GMOs in Brazil.[7] A historical account of the conflict can be found in these works. Some of this material was reexamined in the light of the approach that was incorporated in this study. Documents and materials published by the actors directly linked to the conflict over the release of RR soybean were also examined. Documents where civil society organizations, such as Idec and Greenpeace, present their positions on the subject were examined. Documents linked to governmental bodies such as the Ministry of Science and Technology (MS&T) and CTNBio and documents produced by its former members were consulted, such as letters and statements. Articles from specialized magazines, journalistic articles, and documents from public hearings were also used.

After reading and rereading this material, what can be called the structuring issues of the conflict were defined. Part of these issues can be seen in Tables 4.1, 5.1, and 6.1, which are presented in Chaps. 4, 5, and 6, respectively. With the development of the research and analysis of the material, it was possible to establish a general division in the conflict surrounding the release of RR soybean. Instead of a single conflict, an analysis of the conflict from three distinct axes is offered. These three axes have come to be referred to as the narrative threads of (a) environmental justice (EJ), (b) uncertainty, and (c) labeling. Once these narrative axes and the structuring issues present in each of them were outlined, it was examined how the different actors positioned themselves in relation to the issues present in the conflict. In addition, the examination of national and foreign literature contributed to this analysis, which allowed us to verify in greater detail the elements that make up the frames present in the different discourses that are part of the conflict. The general examination of the position of these actors then made it possible to outline the story line presented in Chaps. 4, 5, and 6.

References

Bleich, E. (2002). Integrating idea into policy-making analysis. Frames and race policies in Britain and Frame. *Comparative Political Studies, 35*(9), 1054–1076.

Cesarino, L. M. C. N. (2006). *Acendendo as luzes da ciência para iluminar as luzes do progresso*. Dissertação (Mestrado em Antropologia Social) – Programa de Pós-Graduação em Antropologia Social. Universidade de Brasília, Brasília.

Cezar, F. G. (2003). *Previsões sobre tecnologias: pressupostos epistemológicos na análise de risco da soja transgênica*. 2003. Dissertação (Mestrado em Filosofia) – Departamento de Filosofia, Universidade de Brasília, Brasília.

---

[7] Among them are the works of Guivant (2005), Pessanha and Wilkinson (2005), Cesarino (2006), Cezar (2003), and Marinho and Marinho (2003).

References

Dryzek, J. S. (2005). *The politics of the earth. Environmental discourses.* Oxford University Press.
Entman, R. M. (1993). Framing. Toward clarification of a fractured paradigm. *Journal of Communication, 43*(4, Autumn).
Fischer, F. (2003). *Reframing public policy: Discursive politics and deliberative practices.* Oxford University Press.
Fischer, F., & Forester, J. (1993). Introduction. In F. Fischer & J. Forester (Eds.), *The argumentative turn in policy analysis and planning.* Duke University Press.
Giddens, A. (1984). *The constitution of society: Outline of the theory of structuration.* Polity Press.
Goffman, E. (1974). *Frames analysis.* Penguin Books.
Gray, B. (1997). Framing and reframing of intractable environmental disputes. In R. Lewicki, R. Bies, & B. Sheppard (Eds.), *Research on negotiation in organizations* (Vol. 6). Emerald.
Guivant, J. S. (2005). A governança dos riscos e os desafios para a redefinição da arena pública no Brasil. In *CGEE – Centro de Gestão e Estudos Estratégicos. Ciência, Tecnologia e Sociedade.* CGEE.
Hajer, M. (1993). Discourse coalitions and the institucionalization of practice: The case of acid rain in Great Britain. In F. Fischer & J. Forester (Eds.), *The argumentative turn in policy analysis and planning.* Duke University Press.
Hajer, M. (1995). *The politics of environmental discourse: Ecological modernization and policy process.* Clarendon Press.
Hajer, M., & Versteeg, W. (2005). A decade of discourse analysis of environmental politics: Achievements, challenges, perspectives. *Journal of Environmental Policy & Planning, 7*(3), 175–184.
Hill, M. (1997). *The policy process in the modern state.* Prentice Hall.
Hill, M., & Ham, C. (1984). *The policy process in the modern capitalist state.* St. Martin's Press.
Juma, C., & Clark, N. (1995). Policy research in Sub-Saharan Africa: An exploration. *Public Administration and Development, 15*, 121–137.
Kaufman, S., & Smith, J. (1999). Framing and reframing in land use change conflicts. *Journal of Architectural and Planning Research, 16*(2), 164–180.
Kaufman, S., Elliot, M., & Shmueli, D. (2003). Frames, framing and reframing. http://crinfo.beyondintractability.org/essay/framing. Acesso em: 15 ago. 2006.
Keeley, J., & Scoones, I. (1999). *Understanding environmental policy processes: A review* (IDS Working Paper, 89). http://www.ids.ac.uk/ids/publicat/wp/wp89.pdf. Acesso em: 1 set. 2006.
Macrae, D. (1993). Guidelines for policy discourse: Consensual versus adversarial. In F. Fischer & J. Forester (Eds.), *The argumentative turn in policy analysis and planning* (pp. 291–318). Duke University Press.
Majone, G. (1989). *Evidence, argument, and persuasion in the policy process.* Yale University Press.
Marinho, C. L., & Marinho, C. (2003). *O discurso polissêmico sobre plantas transgênicas no Brasil: estado da arte.* Tese (Doutorado em Saúde Pública). Escola Nacional de Saúde Pública.
Pessanha, L., & Wilkinson, J. (2005). *Transgênicos, recursos genéticos e segurança alimentar. O que está em jogo nos debates?* Armazém do Ipê.
Purvis, T., & Hunt, A. (1993). Discourse, ideology, discourse, ideology, discourse, ideology. *British Journal of Sociology, 44*(3 (September)), 473–499.
Rein, M., & Schön, D. (1993). Reframing policy discourse. In F. Fischer & J. Forester (Eds.), *The argumentative turn in policy analysis and planning.* Duke University Press.
Rein, M., & Schön, D. (1994). *Frame reflection. Toward the resolution of intractable policy controversies.* Basic Book.
Sabatier, P., & Jenkins-Smith, H. C. (1993a). The advocacy coalition framework: Assessment, revisions, and implications for scholars and practitioners. In P. Saba-Tier & H. C. Jenkins-Smith (Eds.), *Policy change and learning. An advocacy coalition approach.* Westview Press.
Sabatier, P., & Jenkins-Smith, H. C. (1993b). The study of public policy processes. In P. Sabatier & H. C. Jenkins-Smith (Eds.), *Policy change and learning. An advocacy coalition approach.* Westview Press.

Sutton, R. (2006). *The policy process: An overview*. (ODI Working Paper 118). Disponível em: http://www.odi.org.uk/publications/wp118.pdf. Acesso em: 17 ago. 2006.

Triandafyllidou, A., & Fotiou, A. (1998). Sustainability and modernity in the European Union: A frame theory approach to policy-making. *Sociological Research Online, 3*(1).

Weale, A. (1992). *The new politics of pollution*. Manchester University Press.

Wood, L. A., & Kroger, R. O. (2000). *Doing discourse analysis. Methods for studying action in talk and text*. Sage Publications.

Chapter 3
Brave New World of Biotechnology

Ethics, Science, Politics, and Regulation

In the mid-1990s, the debut of GM foods was marked by controversy. Perceived as a great scientific, technological, and economic opportunity by its promoters, its image was overshadowed by the critical voices that emerged in different parts of the planet. The European case, a region that responded with a moratorium that lasted several years, is an emblematic case in this process. The clash between these views has raised several questions. Are GM seeds safe? Should the issue of GM foods be examined only through the lens of environmental safety? Could GM seeds further strengthen the power of the large corporations that produce GM seeds? Could this lead to a loss of autonomy for small farmers? Will biotechnology help promote sustainable agriculture? Are the knowledge and methods used to assess the environmental risks posed by GM foods reliable? Do consumers have a right to know whether the products they are consuming are GM? Are the regulatory policies that governments are putting in place to guide the commercial use of GM foods satisfactory? Or do these policies express a series of errors which should be reversed as soon as possible?

The controversies that occur over GM foods in various places around the world usually cut across issues such as these. If it is true that the controversies produced by emerging technologies are not limited to GM foods, the controversies produced by modern biotechnology tend to attract attention because of the diversity, scope, and intensity of the debates it raises. Hence, Priest notes that "few developments in modern science and technology have engendered as much controversy as the introduction of genetically modified foods" (2005, p. 31\4). These controversies almost always involve the same groups of actors: scientists, farmers, consumers, environmental organizations, public opinion, media, political parties, governments, government agencies, corporations, and political leaders. The configuration of the conflict that occurs among them is not always clear, since it is possible to find polarizations that express different evaluations in each case. Scientists do not have a consensual view on GM foods and so do governments (with their regulatory policies on GMOs) and sectors of the public that express different positions on the issue. The controversies over biotechnology that have developed since its inception are related to the different assessments that are made about the possible adverse effects that may arise from this technology. At the same time, these controversies are influenced by debates about human cloning, stem cell research, etc. At least some studies on GM foods recognize the possible reciprocal influence of these discussions.

It is important here to make an important distinction between scientific controversies and those that occur in public space. From a sociological point of view, a scientific controversy can be defined as "the persistent and public division of members of a scientific community (either individually or as coalitions), who sustain contradictory arguments in the interpretation of reality" (Raynaud, 2015, p. 08). Scientific controversies are not something abnormal in science, but can be seen as an integral element of its own development. As science is inconclusive, causing new theories and discoveries to emerge over time, the clash between new and old beliefs is a natural and predictable process of scientific disputes. For Bridgstock (1998), purely scientific controversies are those in which only scientists take part. In general, scientific controversies revolve around three fundamental issues: facts, theories, and principles.

Scientific controversies can be distinguished from public controversies. The latter are captured by the idea of public or social problem. Public controversies can be defined as "debates about a current issue of public interest that comprise more or less important social stakes in a given culture" (Amossy, 2021, 32). Scientific controversies tend to operate within the scientific field and are driven by scientific positions held by scientists themselves. Public controversies, in turn, may be influenced by scientific controversies, as is the case with GM foods, but they are not limited to the scientific field. On the contrary, they can invade the most different spheres of social life (e.g., religion, culture, politics). This allows scientific controversies to occur in a more organized way, following the very rules of scientific debate, while public controversies tend to occur in a less organized way. The latter may include a set of concerns that are usually disregarded by scientists. And, in general, it includes a much greater diversity of debaters.

This intimate and almost direct relationship between scientific and public controversy occurs due to the increasing technical applicability of scientific knowledge in society. Genetic discoveries soon become techniques for the manipulation of living beings that, through the market, can bring direct implications to people's daily lives. Scientific controversies are not always isolated in the scientific field, but may have implications for people's lives even before the scientific community establishes any consensus on the issue under debate. In the case of biotechnology, it can have direct implications for the relationship people have with their eating practices. Thus, if an individual is suspicious of the safety of GM foods, he or she will simultaneously face a set of uncertainties associated with these practices. Therefore, as Markett and Backer point out, as "biotechnology affects us more in our lives, it will be subject to more controversy and discussion" (2003, p. 21). If scientific discoveries have more and more influence on people's daily lives, in processes involving the consumption or production of food (GMOs), then scientific controversies can hardly remain circumscribed to the scientific field, since the usefulness of this knowledge is measured by the changes it operates in people's lives.

In this way, the facts of science can be criticized in the context of social life, appealing, for this, to science itself. In modernity characterized by high reflexivity of people and institutions, as Giddens says, "few can afford to ignore claimed findings in relation to, for example, the benefits and risks of eating various kinds of foods, health hazards of various sorts, ecological hazards and so forth" (1994, 216). One consequence of this is that if in the past scientific and public debates were kept relatively separate, today this separation seems less likely to occur. For if it is true that science has increasingly practical implications for people's lives, it is possible that people will examine this influence in an increasingly reflexive way:

> Many scientific claims enter public discussion well before – according to the pre-established formulae of science – they could be said to be 'proven'. Where the 'bads' of scientific innovations are concerned, individuals or groups are often disinclined to wait until claims have been 'properly established', for the dangers they may face if these claims are valid may be pressing. Hence it is not surprising that, for instance, food scares become common. A particular food item that one is used to consuming in a regular way suddenly becomes suspect, no matter what the scientists – or some of them, given that scientific opinion is frequently divided – say on the issue. (Giddens, 1994, 216)

Public controversies tend to have three important characteristics. They involve (a) conflicts over beliefs, values, and interests that transcend the simple pursuit of greater knowledge (science). They tend to arise because of (b) scientific risks and uncertainties that are perceived by the public to exist. These uncertainties are generally associated with the risks associated with the application of new technologies. And also by the fact that specific interests of social groups, beyond the scientists themselves, project themselves into public discourse (National Academies of Sciences, Engineering, and Medicine, 2017, 51). Public controversies tend to be linked to disputes involving beliefs and values, which may involve "moral and ethical considerations – all of which are central to decisions and typically subject to public debate" (National Academies of Sciences, Engineering, and Medicine, 2017, 53).[1] At the same time, because scientific controversies involve "uncertainty, and concerns about risk and the consequences" (National Academies of Sciences,

---

[1] The report of the National Academics of Sciences, Engineering, and Medicine (2017, 54) states that organized interests often seek to protect their interests by promoting the polarization that a controversy tends to produce. But they also point out that, in addition to the citizens themselves, the scientists themselves and their organizations also have economic, professional, and personal interests in these disputes, which in turn can influence the trust that the public places in expert discourse, that can influence the focus of their communication efforts: "scientists and their organizations also have economic, professional, and personal interests that they may defend and promote, a fact that may influence the focus of their communication efforts."

Engineering, and Medicine, 2017), they can expand the number and diversity of people involved in public conflict. The National Academies of Sciences, then, notes that government officials may try to draw on scientific knowledge or scientific experts "to help craft solutions to policy issues of concern to the public" (National Academies of Sciences, Engineering, and Medicine, 2017, p. 63). Judges and courts may do the same to render a verdict for a particular case. The media themselves may attempt to draw on scientific knowledge to speak on behalf of their audience's interests. In short, the most different actors and groups involved in the debate may use science that supports their claims. Thus, they note even scientists can be seen as stakeholders with their own motivations and interests at stake. Public controversies tend to strongly influence technological development, because they directly influence the public acceptance of the innovations that arise from it. And this acceptance depends on several factors: utility, ethical implications, distributional issues, power, etc. Hence, controversies surrounding biotechnology are currently perceived as a learning opportunity to rethink the development of other types of emerging technologies such as nanotechnology.[2]

3.1 The Radical Nature of Biotechnology

In his book *The uses of life. A history of biotechnology*, Robert Bud locates the origin of the word *biotechnology* in the writings of Hungarian agricultural engineer Karl Ereky. Ereky, who went on to serve as minister of nutrition in his country, used the new word as part of a "campaign intended to supersede the backward peasant" (Bud, 1993, p. 32). Between the years 1917 and 1919, Ereky will write three texts where he will make use of the new word. In these texts he points out that not only the organization of agricultural labor could be approached in a scientific manner, in Fordist terms, but the animals themselves could be approached from such a perspective. As Bud (1993, 34) will note, for Ereky "the pig was a machine converting carefully calculated amounts of input into meat output. Indeed, he described the pig as a "Biotechnologische Arbeitsmaschine"".[3]

Existing texts in literature do not always select the same events and dates to demarcate the emergence of modern biotechnology. That is because the factors that helped structure biotechnology did not emerge simultaneously. Genetic manipulation techniques that are associated with biotechnology emerged after the 1950s when James Watson and Frances Crick discovered the structure of DNA. And the

---

[2] For an analysis of the possible lessons that the surrounding debates about biotechnology can offer for nanotechnology, see Kenneth and Thompson (2008).

[3] Krimsky and Gruber (2014) observe that in the vision of modern agricultural Baconians, "farms are like factories" and "food production must be as efficient as an assembly line." This means that in the discourse of the most enthusiastic promoters of biotechnology, this technology is seen as allowing "to speed up food production to cultivate more crops per given acre, per unit of time, per unit of labor, and per unit of resource input."

3.1 The Radical Nature of Biotechnology

biotechnology industry had only taken its first steps in the mid-1970s. Regulatory policies aimed at the use and marketing of GM foods began to emerge in the 1980s. And although the commercial release of GM products occurred in the late 1980s, the public controversies that gained media attention emerged in the 1990s.

Modern biotechnology promises us a brave new world (Priest, 2005). From sheep that produce pharmaceutical drugs through their milk, it also promises us GM pigs whose organs can be transplanted into humans. A series of other promises could be mentioned here. These and other promises are part of the socio-technical imaginary that integrates the discourse of modern biotechnology (Jasanoff & Kim, 2015). In this "brave new world of biotechnology," a number of economic, environmental, and social opportunities are seen as achievable without major upheavals. If biotechnology is, on the one hand, presented as a major scientific, technological and economic revolution, on the other, it is simultaneously perceived by its promoters as equivalent to conventional technologies when it comes to risk. In a utilitarian's view, biotechnology would offer the best of all worlds: the possibility of maximizing opportunities in the most different areas of life without simultaneously generating any additional loss.

It may be important here to take a brief look at the commonly presented definition of biotechnology because of its implications for the debate about its social and environmental implications. Thus, it is not uncommon for promoters of biotechnology to almost always remind us of this trivial fact: that biotechnology has been practiced throughout human history and that selection of plants and animals has taken place over this period. This association is made in order to bring biotechnology closer to a practice that is considered normal, ancient, or conventional. This can eventually produce a sense of familiarity and security. By establishing a line of continuity between present and past practices, one tends to do the same with their possible adverse effects. Therefore, for biotechnology's enthusiastic promoters, the risks brought by it would not be substantially different from those existing in conventional animal and plant breeding, an argument that, by the way, usually emerges in conflicts associated with GM foods and that, as will be examined in the next chapters, also tends to appear in the Brazilian case involving the release of RR soybean.

Human societies have always manipulated plants, animals, and other organisms. Much of what we eat has been altered, modified, or improved by human beings throughout their history on the planet (McHughen, 2008, p. 36). Examined from this perspective, biotechnology could be seen from a continuum of techniques applied in the manipulation of living organisms over the course of this long journey. Therefore, if one takes this perspective, the history of biotechnology could not be told in decades but in millennia. In some definitions that are offered of biotechnology in the scientific literature, it is this vision that will be found, a view that puts biotechnology in rather generic terms. One such view is offered by Markert and Backer (2003, p. 20) when they define biotechnology as the "manipulation of biological organisms to make products benefiting humans." At the same time, they state that the practice of biotechnology has existed for a thousand years (Markert & Backer, 2003, p. 20). However, they recognize the difference between traditional and modern biotechnology. But this definition is not as useful as one might imagine, because as McHughen (2008, p. 36) will say "if we define biotechnology as the application of science or technology to biological systems, as many do, then biotechnology started 10,000 years ago." When examining a textbook that introduces the subject, one finds the recognition of a rupture produced by current biotechnological practices when compared to plant and animal selection practices in the past. Thus, authors such as Thieman and Palladino state, in a book with this profile, that the transformation of plants provided by biotechnology "can achieve results that are impossible with conventional breeding methods" (Thieman & Palladino, 2020, p. 190). One of the results that are possible with current genetic manipulation, and that would be impossible with the techniques of the past, is the "engineering a crop plant to be resistant to herbicides" (Thieman & Palladino, 2020, p. 190)[4].

The new biotechnology cannot be perceived as something equivalent to older plant and animal selection practices because it now allows one to achieve results that would be impossible with the methods of the past. If it is possible to perceive continuities between today's biotechnology and the one of the past, it is difficult to escape the environmental, social, political, economic, and ethical implications of this rupture. This difference tends to be important because the current controversies surrounding biotechnology tend to arise precisely from "attempts to directly manipulate DNA in a laboratory and because genetic material and traits from one species are being introduced into another species" (James, 2018a, b). Thus, when political leaders, scientists, and citizens show concern about current biotechnology, they are not necessarily directing their concerns towards the artisanal production of things like cheese or beer. These concerns arise from the application of modern genetic engineering that allows direct manipulation, using laboratory techniques, of the genetic information of living organisms. So, for critics, that is the difference that makes the difference.

The deeper the knowledge offered by science, the greater the level of manipulation of the world that occurs through the technology that emerges from that same knowledge. Thus, when compared to conventional methods of plant and animal selection, biotechnology represents a deeper level of knowledge (DNA) and, consequently, of manipulation of living beings. By creating conditions for a new set of techniques to manipulate living organisms, modern biotechnology also creates the possibility of creating new biotic artifacts:

> Finally, with the arrival of DNA genetics and biology at an even deeper level of understanding, the technology it engenders operates at a correspondingly far deeper level, namely the

---

[4] This distinction is not only recognized by scientists working directly with modern biotechnology but is also found in government documents that seek to create regulatory policies for modern biotechnology. These new guidelines are concerned with the deliberate and direct alteration made possible by new techniques of genetic manipulation of living things. As Kim informs us, "The terminology for the application of biotechnology to food and agriculture has been used in various forms, such as genetic modification (GM), genetic engineering (GE), genetic manipulation, gene technology, and rDNA technology. However, the collective term 'genetically modified organisms (GMOs)' is used frequently in regulatory documents and in scientific literature to describe the deliberate introduction of DNA by human intervention into plants, animals, and microorganisms" (2014, 153).

3.1 The Radical Nature of Biotechnology

molecular level, thereby generating an even greater level of artifacticity in its end products. Biotechnology no longer relies on breeding in the traditional sense which underpins both craft-based and Mendelian technologies, but bypasses it; (...) This testifies to the radical nature of biotechnology, the deep level at which it manipulates genetic material, and hence the depth of artifacticity in its transgenic products. (LEE, 2009, p. 102)

By lending its theoretical knowledge to the techniques of genetic manipulation, the new biotechnology has not only entailed changing techniques for genetically altering living things, but also the ontological transformation of living beings to become biotic artifacts (LEE, 2009, 103). Therefore, as a consequence of this change, not only are biotechnological techniques today distinct, but the products generated by them are ontologically different. The new means produce new biotic artifacts that, as Lee (2009) points out, show us the radical nature of biotechnology.

Biotechnology is a word that refers to the different techniques of genetic manipulation of living beings. GM soybean, whose economic release generated a series of conflicts at the end of the 1990s in Brazil, is an innovation brought about, for example, by agricultural biotechnology. But agricultural biotechnology is only one type of biotechnology existing today. Besides it, you can also find microbial biotechnology, animal biotechnology, forensic biotechnology, bioremediation, aquatic biotechnology, and medical biotechnology. Biotechnology can thus be confused with other more general terms that can be found in the literature that embrace issues linked to genetics. These terms include gene manipulation, gene cloning, recombinant DNA technology, genetic modification, and genetics. Although some of these processes are related, for the most part they refer to distinct things. For the purposes of the discussion that will be made in this and the following chapters, biotechnology can be defined as "any of several techniques used to add, delete, or amend genetic information in a plant, animal or microbe" (McHughen, 2008, p. 36). Or, in the terms that Priest (2005) presents:

Currently the term most often refers specifically to active manipulation of the genetic code of life, the deoxyribose nucleic acid (DNA) molecule – sometimes resulting in recombinant DNA (rDNA) products in which the genes of one species are combined with those of another to produce a desired trait (genetic engineering). (Priest, 2005, p. 34)

Modern biotechnology makes it possible to transfer genes from one organism to another without respecting species boundaries. It thus uses the technical capabilities offered by genetic engineering. Genetic engineering has been defined as the "intentional transfer of genetic material from one organism to another, usually of a different species" (Reiss, 2001, p. 13). Genetic engineering, therefore, is a technique used in many different types of biotechnology. It allows the transfer of genetic information from one organism to another while enabling research with stem cells, cloning among other things. Biotechnology is an umbrella word that embraces all these techniques (Markert & Backer, 2003, p. 21).

In the following part, some issues that have been part of the controversies about the commercialization of GM foods are examined. Controversies most closely associated with biotechnology. These issues are examined because they shed light on the conflict over the liberalization of the commercialization of GM soybean in Brazil, which will be analyzed in the following chapters. Some of the issues examined below will be taken up again throughout the book as the conflict in Brazil is addressed.

3.2 The Risks of Biotechnology

For the most enthusiastic supporters of biotechnology, if the first green revolution occurred through the use of pesticides in the agricultural system, the second will be promoted with GM seeds. For this group, biotech represents the opportunity for a great productive leap in agriculture through the use of GM seeds. In places where agriculture is not feasible today, due to different environmental conditions, it could become possible now with the new technology. And where it already exists, a leap in productivity could also be promoted in order to increase existing productivity. Allied to this economic efficiency, biotechnology is seen as bringing ecological benefits. Thus, the increase in agricultural productivity could occur simultaneously with the reduction of water and fertilizer consumption in agriculture that, in turn, would produce free positive environmental effects. Among them would be the reduction of deforestation. With the gain in agricultural productivity, the pressure to cut down forests in order to make the soil useful for plantations could be reduced. If that were not enough, some also see biotechnology as a solution to world hunger.[5] Less use of water resources, greater agricultural productivity, less pressure for deforestation, and an answer to world hunger: too good to be true? Biotechnology's most enthusiastic advocates would say the brave new world is now possible. In this scenario, biotechnology could be seen as a great ally to combat global warming and promote a more just world by creating the conditions for global food security. If sustainable development is development that is environmentally sustainable and socially just, biotechnology could be seen here as one of its main foundations in the techno-scientific field.[6]

However, since the emergence of GM foods on the market, more optimistic arguments about the new technology have emerged competing with more critical views that seek to cast a shadow over these expectations. Critics draw attention to its possible perverse effects, which generally take the form of many different types of risk (environmental, economic, social, etc.). In all of these cases, if biotechnology points to benefits of some sort, it is seen as also implying some sort of danger. For many of those who see such dangers as possible, whether in the present or the future, biotechnology could produce the opposite result that is suggested by its most

---

[5] On the agbiotech debate and the issue of food security in the world, see Altieri and Rosset (2002).

[6] In the European case, until the mid-1990s, agbiotech was seen from the perspective of ecological modernization which, like certain interpretations associated with the concept of sustainable development, tends to link agbiotech to the values of environmental sustainability. For the approach to agricultural biotechnology from the perspective of ecological modernization, see Levidow's critical analysis (2014a, b)

enthusiastic advocates. In that case, biotechnology could generate significant or even catastrophic environmental impacts, destroy agricultural livelihoods, lead to greater concentration of wealth, and more. At the same time, the different criticisms that can be made of biotechnology can present themselves on different levels. Critics of biotechnology may differ widely in their assessments, and rather than an outright rejection of the technology, they may call for different laws or standards to regulate its use. In this case, the dispute is not always about acceptance or rejection of the technology, but may involve different views on how best to conduct it in the regulatory process.

In the evaluation of Leiss and Chociolko (1994), it is the risks and uncertainties that are at the origin of public controversies about techno-scientific development. In their book *Risk and Responsibility*, the authors note that while the public is generally familiar with practices that involve some risk, they tend to express concern about those that affect health or conditions in the natural environment (Leiss & Chociolko, 1994, p. 04). As they note, there is a rather simple reason why the public eventually expresses these concerns. They arise because of "fear of falling victim unfairly to uncompensated loss" (Leiss & Chociolko, 1994, p. 04). For Leiss and Schociollo (1994, p. 06), risk means exposure to the chance of loss. It is these possibilities of loss, perceived as real or potential, that underlie public controversies about the risks of GM foods. Uncertainties, therefore, tend to matter to the extent that the public acknowledges the existence of a risk. At the same time, public apprehensions about these risks also tend to emerge from the fact that people consider that their possible exposure is deliberately induced by some social actor in order to obtain some benefit (Leiss & Chociolko, 1994, p. 06). Therefore, they argue that the risks associated with new technologies tend to generate controversy because the public perceives that they cannot control their exposure to the possibility of unjust losses (Leiss & Chociolko, 1994, p. 08).[7] This tends to occur because risk, in many situations, is not the result of an individual decision, but emerges for individuals, families, and communities as a result of decisions made by other agents. Many risks that people face arise from the fact that they are intentionally produced by others. Hence, risk assessments, regardless of who makes them, can have important distributional effects. Risk assessments, and the policies guided by those assessments, then determine the distribution of opportunities and costs associated with the decisions that produced them. And the dangers that certain risk decisions assume do not always fall on those responsible for those decisions.

It is not uncommon to find underlying by those who propose technological innovation a utilitarian perspective of evaluation of existing risks. A utilitarian ethical perspective "holds that we ought always to act so that we maximize good consequences and minimize harmful consequences" (Comstock, 2002, p. 91). The

---

[7] Among the losses that would be of concern to the public, indicated by Leiss and Chociolko (1994), would be losses associated with such things as (a) physical consequences (e.g., death or ill health), (b) possible adverse psychological effects (e.g., stress), (c) compromised livelihood, (d) loss of wealth or personal property (e.g., home), and (e) loss of future opportunities associated with career or personal income.

utilitarian view is not a perspective restricted to utilitarian philosophers, but can be seen as present in the analyses generally used to evaluate the impact of technological innovations. This is because, in many cases, the analyses of these innovations occur through the application of cost and benefit analysis (CBA). As Dusek (2006, p. 56) will observe, a "risk/benefit analyst evaluate the worthwhileness of technological projects by adding benefits and subtracting risk in a manner similar to Bentham's utilitarianism." Thus, in certain cases, what advocates of new technologies including biotechnology suggest is not that they are entirely risk-free, but that their benefits tend to outweigh them in some way. Which in turn should make us then accept the technology. Since the benefits are greater than the costs (risks), the acceptance of technology should constitute a predictable and rational behavior. Critics of this utilitarian view not only seek to point out possible risks that can be neglected, but also the mistakes that may be underlying the attempt to examine, through a utilitarian viewpoint, the existing risks. This is because utilitarian calculation could make some sense in situations where gains and losses can be calculated and measured by a common denominator (economic value). But it is not certain that gains and losses can be calculated in this way. This may suggest, then, the inadequacy of trying to evaluate new technologies through these assumptions.

Although many controversies arise over the environmental implications brought about by biotechnology, existing controversies are not always reduced to this issue. This is because, in many cases, concerns are directed at the implications biotechnology may have for other issues. Critics of biotechnology argue that it can make existing economic relations in agricultural systems even more asymmetrical. It could also have negative implications for food security, small farmers, animal welfare, etc. The resistance to GM foods that occurred in Europe in the late 1990s was associated with many issues that transcend mere concern about the environmental risk posed by biotechnology. These conflicts were also associated with policies seeking to promote different agricultural models in Europe. Therefore, James will observe that:

> critics of agricultural biotechnology are not singularly concerned about food and crop safety. Intellectual property rights, the adequacy of regulations, labeling and corporate power over the development and use of GM technologies are examples of important, relevant and unresolved problems. Proponents of agricultural biotechnology who push safety claims when opponents argue about inadequate regulation or corporate power miss this point. Furthermore, one of the main arguments proponents of biotechnology make is that it is needed to feed the world (...) . This argument might be relevant in the case of some developing countries, but it is not necessarily relevant in developed countries where food availability is less of a problem than other food security concerns, such as access and adequacy. (James, 2018a, p. 02)

In short, the debate on the consequences of biotechnology is not reduced to its environmental risks, but encompasses a wider range of issues. At the same time, these issues may become more or less relevant depending on the cultural context in which they are considered. Issues associated with property rights, labeling systems, and corporate power are just a few of the questions that still trouble public opinion. And none of them can be reduced to issues of environmental safety.

3.3 The Ethics of GM Food

The ethical issues brought about by biotechnology are not, for authors such as Thompson (2002), entirely new. In fact, biotechnology reconfigures, he argues, many of the issues already being examined with the impact of other types of technology on agriculture.[8] Thompson (2002) traces the origin of the controversies over the ethical implications of genetics to the 1975 Asilomar Conference. In the specific case of biotechnology, the author locates the emergence of this debate with the lawsuit filed by Jeremy Rifkin in 1984, which forced the National Institutes of Health's Recombinant DNA Advisory Committee (the RAC) to consider the environmental impacts associated with the creation of ice-nucleating bacteria to protect strawberry crops from low temperatures. In his assessment of the ethical issues associated with biotechnology, Thompson (2002) finds four main areas through which these issues tend to arise. The cases offered by the author will be used to illustrate some of the ethical implications brought by biotechnology.[9]

In the context of food safety, the ethical dispute occurs between a scientific risk assessment oriented towards food safety and another position founded on consumer choice. The first position assumes that the risks of GM foods can be examined scientifically. In this approach, the position of consumers to assess risks and make choices from this view tends to be seen as inappropriate. Those who advocate consumer choice argue that, regardless of what the risk assessment may indicate, the ultimate decision about the consumption of GM foods should be left to the consumer themselves (Thompson, 2002, p. 69). In short, even if GM foods are considered safe, it is up to the consumer to decide whether to consume them. It should be noted that, from this perspective, the issue of the environmental safety of GM foods does not allow for a simple and definitive resolution of the dispute because, as Thompson points out: "This is an ethical issue rather than a simple dispute over facts because one viewpoint stresses individual autonomy and consent, while the other stresses rational optimization" (Thompson, 2002, p. 70). This type of conflict will appear, as can be seen, in the controversies that arose in Brazil in the debate on the labeling of RR soybean.

A second set of ethical disputes are associated with the possible environmental risks associated with biotechnology, an issue that will also emerge in the conflict over the commercial release of RR soybean in Brazil. Critics point out that such risks exist and, in part, are unacceptable. GM food advocates, on the other hand, consider that scientific assessments of the environmental risks of biotechnology are

---

[8] The ethical debates that have developed about biotechnology over the past three decades could have been prompted, according to Thompson (2002), by other types of technologies that provoke similar concerns. After all, issues associated with food security, impact on environment, and smallholder autonomy are not concerns that remain restricted to the agbiotech debate. Thus, for Thompson (2002, p. 69), the debate on "agricultural biotechnology is (...) a surrogate for debate over technological progress itself."

[9] Other works that examine the ethical implications of biotechnology are Robinson (1999), Thompson (1997), and James (2018a, b).

reliable and the data emerging from such research and assessments authorize the commercial release of GM foods. But the debate taking place about the possible environmental risks of GM foods is not, for Thompson (2002, p. 70) a debate that occurs in strictly factual terms:

> Like debates over food safety, they involve disputes over the validity and wisdom of relying on off-setting cost-risk-benefit optimization to conceptualize the issues. Even among those who accept the risk-benefit approach, the issues involve value judgements about the relative importance of food production as opposed to the preservation of wildlife and genetic diversity. They involve value judgment about how to proceed in the face of uncertainty, and indeed, about the very nature of uncertainty. The issues involve value judgments even about the nature of nature, as some believe that preserving wildlife and a certain aesthetic character on farms is part of nature conservation, whereas others see agriculture as inimical to wild nature. (Thompson, 2002)

Imagine that certain GM foods are safe for human consumption, but that they result in significant risks to wilderness areas. Would a cost–benefit analysis take these impacts into account? Within the scope of approaches that seek to evaluate technological innovation, Dusek reminds us that cost/benefit analysis has great difficulty in evaluating in monetary terms the loss of wildlife or noncommercial species of living beings. This is because, according to him, "if one takes the simplest approach, endangered species or living things with no commercial use simply have zero value and their loss counts for nothing. If one takes the most simple and straightforward approach to the commercial value of wildlife, their value may be very low" (Dusek, 2006, p. 64). Then, when examining the European Parliament Directive (EC) 2004/35, Bodiguel and Cardwell (2010, p. 17) call attention to the fact that in this document "environmental damage is defined by reference to protected species and natural habitats; and GMOs are more likely to be grown in area without conservation status." This means that areas that have the legal conservation status tend to be considered in the impact assessment, while those that do not have this status may escape any risk assessment. This may be perceived as natural to technocrats, but to the public it may be something that should be open to moral and political debate. Therefore, one should not ignore the fact that, in some cases, it is the public, not RA, that can broaden the debate on these issues.

A third area of dispute is associated with the social consequences produced by GM foods. Advocates argue that biotechnology can contribute to solving world hunger and strengthening agriculture in regions that are unsuitable for growing crops. The debate over benefits to the poor would give biotechnology the potential to promote distributional outcomes. Given that inequality, hunger, and food insecurity are seen as morally undesirable, biotechnology's effects would make it an ally in global distributive politics where the poorest would win. However, critics, Thompson (2002, p. 71) reminds us, suggest that biotechnology could disable traditional agricultural ecosystems, impoverish farming communities, and lead to a concentration of land and wealth in the most vulnerable rural areas. For critics, the property system that accompanies biotechnology is not in the interest of the poorest farmers and will not even serve to raise their standard of living. Therefore, it must be considered that biotechnology is embedded in a property system that may work unfavorably for smallholders. Again, debates about the social consequences of GMOs in agriculture would operate, for Thompson (2002) within the framework of distinct ethical values. "There are, thus, ethically grounded notions of fairness and distributive justice lurking in debates over social consequences" (Thompson, 2002, p. 71).

Finally, Thompson (2002, p. 69) views the ethical issues associated with the trust that is placed in biotechnology itself and the institutions that support its development. But this is an issue that will be examined below, when considering the question of public participation, as these issues can be seen as related to each other.

3.4 Agricultural Innovation and Distributives Issues

In his book *Social Justice and Agricultural Innovation*, Timmermann (2020) seeks to examine the complexity surrounding the relationship between food systems and social justice. In this work, the author reminds us that food production systems operate in the midst of a complex set of social, ethical, political, economic, and environmental factors. The changes brought about by these systems affect, he says:

> social and environmental interactions and thus require innovation systems to continuously (re)align with the demands of social justice. Food procurement is a major social justice challenge, as constantly transforming food systems have to be continuously assessed and readjusted to meet the demands of justice in all its dimensions. Current as well as future food needs raise issues of justice. (Timmermann, 2020, p. 01)

Many of the current discussions of food production systems raise questions of this kind. For example, agricultural biotechnology can directly interfere with our diets and therefore can lead to eating disorders that can produce different effects on the health of a population ranging from malnutrition to obesity. Therefore, one tends to agree with the statement that a "food production systems should not lead to avoidable health hazards" (Timmermann, 2020, p. 02). Within the sphere of labor, issues of justice can also be seen to cut across the biotechnology debate. Thus, one can agree with Timmermann's (2020, p. 02) statement that food systems should be "reformed in order to be socially sustainable in terms of protecting farm workers' health and rights, and provide decent standards of living in rural areas." Or consider the issue of small farmers. Timmermann (2020, p. 02) also states that agricultural food systems should not cause small farmers to massively and abruptly abandon rural areas. In addition, the ethical implications associated with commitments to future generations can be considered depending on how these farming systems are created. Commitment to justice for future generations should lead us to examine the environmental impacts produced by current food systems, given that these systems have enormous environmental impacts. The relationship between agricultural biotechnology and social justice, then, is a close one. Understandably, these issues also enter the GM food debate in some way. After all, one of the arguments generally used by advocates of biotechnology, as seen above, is that it would help combat hunger and food insecurity in the world. Thus, GM foods would be a way to promote a more just and secure global agricultural order. However, while the argument for agricultural biotechnology may reference social and EJ issues to give it greater legitimacy, others may use opposing arguments to criticize it. Thus, agricultural biotechnology's implications for social and EJ encompass controversies about its social impact as well.

3.5 Public Participation for What?

Public participation has been seen as important in the formulation of public policies for GM food in several places. This premise can be found in key documents that seek to provide guidance for contemporary environmental policy. At the international level, there are several initiatives that point to the need for public participation in deliberations about biotechnology. In this area, the *Convention on Biological Diversity*, *Cartagena Protocol*, *Aarhus Convention*, and, perhaps less known in the literature, the *Organization of African Unity* stand out. Although the *Convention on Biological Diversity* has been little considered in the GMO debate, it pioneered the need for public participation in the GMO regulatory process.[10] The *Cartagena Protocol* will offer similar guidance when it states in article 23 that decisions to be taken involving GMOs "shall make the results of such decisions available to the public" (United Nations, 2000, p. 18). The *Aarhus Convention* will reaffirm these commitments, recognizing "the concern of the public about the deliberate release of genetically modified organisms into the environment and the need for increased transparency and greater public participation in decision-making in this field" (UNECE, 1998, p. 03).[11]

Given the prescription that these documents offer on the issue, then, it is worth asking: After all, why would public participation be important in the decision-making process involving GM foods? Public participation is important because, as Bodiguel and Cardwell (2010) point out, GM foods are intended to promote the public's well-being. It is the public who will ultimately consume these products. Therefore, it would not be strange to consider that this same public could influence decisions about what it consumes. At the same time, there is research evidence to

---

[10] Bodiguel and Cardwell (2010) make reference to article 14 of the Biological Convention where the following guidelines can be found: "Introduce appropriate procedures requiring environmental impact assessment of its proposed projects that are likely to have significant adverse effects on biological diversity with a view to avoiding or minimizing such effects and, where appropriate, allow for public participation in such procedures" (United Nations, 1992, 09).

[11] As Bodiguel and Cardwell (2010) point out, when comparing the Cartagena Protocol with the Convention on Biological Diversity, one can see that public participation is less precise in the first case. The obligation for public participation in this document is presented as an important element of the decision-making process, but not as a necessary condition for it. According to the authors, the obligation of public participation "seems to be an obligation only to consult, not an obligation to feed the responses into the decision-making process" (Bodiguel & Cardwell, 2010, p. 14).

3.5 Public Participation for What?

indicate that the public tends to support mandatory labeling for GMOs. However, if GM foods are perceived to be safe and the scientific knowledge authorizing this assessment is considered to be entirely reliable, the need for public participation may be seen as unnecessary. In addition to potentially inducing public restriction, this regulatory model tends to strongly influence the risk communication of government policy. This model is founded on the thesis of the existence of a knowledge deficit on the part of the public (Thayill, 2014, p. 11). This model assumes that the resistances that the public expresses to GM foods originate in a subjectively produced irrationality. This perspective expresses a technocratic view of risk communication. It assumes that the public tends to be incapable of understanding the risk issues associated with GM foods. However, as seen in the previous part, if scientists themselves can disagree about the safety of GM foods, why should it be doubted that the public has access to this information? And, at the same time, what would prevent the public itself from using these criticisms to guide its judgment about GM foods? And if the disagreement would be a reflection of irrationality, what about the disagreement expressed by scientists themselves? What of the GM food resistance groups who use this scientific information to validate their position in the public debate? The consequence of this regulatory model is that it will tend to create simplistic risk communication policies that feed erroneous assumptions about human behavior and public reaction to biotechnology.[12]

If controversies over GM foods tend to involve ethical issues that transcend a mere concern for the environmental safety of these products, the technocratic model based on the knowledge deficit tends to prove somewhat deaf to the demands of the public. The issue of trust brings us back to the importance of public participation for the regulation of biotechnology because, if the regulatory process is tasked with addressing these very issues, then a greater dialogue between regulators and the public should be established. Which in turn seems to require greater public participation in the policy process that will define the use of GM foods. Public acceptance of GM foods is related to the ethical issues mentioned above. And, as noted, among the ethical issues associated with the release of GM foods for use is the question of public trust in science:

> Even in an explicitly ethical mode, the question of trust inevitably connects with the border public's attitudes and perceptions of biotechnology. My suggestion today is that the way that researchers, regulators, and administrators comport themselves with respect to the ethical issues I have already reviewed, albeit briefly, is the largest single factor in determining whether they are trustworthy. (Thompson, 2002, p. 72)

---

[12] Research shows that educating the public is important to the regulatory process of GMOs; however, other work also shows that when public education occurs on a strictly scientific basis, it does not automatically tend to produce greater popularity for GMOs (Bodiguel & Cardwell, 2010, p. 10). The trust that lay people place in government and experts is an important factor in how they assess the risks associated with the technology. Public distrust is also associated with situations where the government and the experts themselves have been viewed as suspicious by the public for some reason. It is important to note that the increasing flow of information about GM foods has not substantially altered the public's perception of these products over time. On this point, see Thayyil (2014, p. 07).

It is not at all obvious that the disputes taking place around GM foods are then confined to the environmental safety of biotechnology. Nor, at the same time, does the public have confidence in science with only this issue in mind. They may be dissatisfied with how the institutions responsible for regulating biotechnology end up addressing a larger set of issues that escape the usual RA. Summarizing Thompson's argument (2002), the situation can be considered as follows: the inability of responsible institutions to address the different ethical issues of concern to the public is one of the factors feeding the public's distrust of biotechnology. This distrust may in turn fuel resistance to biotechnology. And public rejection, in turn, does not translate to a rejection of the technology itself, but rather of the way it is implemented and promoted. At least, this is an assumption that should be taken with some degree of seriousness.

The consideration of the role of the public in the regulatory process should also lead us to a review of the role of contestation in the development of new technologies. As Thayyil (2014, p. 05) indicates, contestation, more than an obstacle, can signal real or fictional fears that, regardless of their origin, need to be considered. The contestation against technological innovations can also be important for the process of democratization of the technology itself. Public participation is associated with how law and regulation incorporate scientific knowledge and how they apply this knowledge in the regulatory process (Thayyil, 2014, p. 05).[13] And these processes tend to influence how the technical-scientific development operates within society itself.

3.6 Regulating GMOs

Regulation is generally seen as expressing a type of government activity that is exercised by public agencies that, when performing their functions, restrict the behavior of members of a political community in order to prevent undesirable outcomes (Baldwin et al., 2012, pp. 2–3). It is sometimes seen as a (a) set of commands implemented by the government via its agencies or organizations. It is also seen as a (b) type of deliberate influence exerted by the state. Or it can be seen more broadly as (c) all forms of social and economic influence that can be exercised by the state but also by the market and even civil society. In the first case, regulation is seen as involving the promulgation of a set of rules to be enforced by state agencies. In the second case, regulation has a broader meaning to the extent that its meaning covers all possible actions to be taken by the state to promote this influence on society. And

---

[13] One example where government policy sought to promote greater public engagement was in the UK. The public debate on GMOs took place in the UK in 2002–2003 and has been seen as an unprecedented experiment in engaging the public in the process of formulating public policy. For an analysis of the public participation process that took place in the UK, see the work of Horlick-Jones et al. (2007).

3.6 Regulating GMOs

the third meaning is even broader in that it suggests that regulation can be exercised beyond the government itself.[14]

There are several bodies and institutions that establish policies and protocols for the use and marketing of GM foods internationally. This regulation occurs through the Codex Alimentarius Commission (CAC), the Organization for Economic Cooperation and Development (OECD), and the Cartagena Protocol on Biosafety (CPB) (Kim, 2014). The CAC is one of them and its policies work together with FAO/WHO in order to establish guidelines under the food safety policy. The purpose of the CAC is to protect the health of consumers, ensure fair practices within the food trade, and promote coordination of food standards. According to Kim (2014, 162), the "Codex's objectives include consideration of standards, guidelines, or other recommendations as appropriate for foods derived from biotechnology or traits introduced into foods by biotechnology on the basis of scientific evidence and risk analysis." The principles include a scientific and premarket risk assessment and an evaluation of direct unintended effects that may arise from the insertion of new genes into plants and foods. The OECD, for its part, has provided policy guidance on biosafety for feed and food marketing. In addition to stating that its aim is to protect the environment, one of its main objectives is to reduce unnecessary barriers to trade (Kim, 2014, p. 163). Associated issues are also found in the Cartagena Protocol on Biosafety.

While these international initiatives seek to establish a consensus on the regulatory policy for GM foods on an international scale, it is possible to say that this effort is faced with great polarization in the scope of countries' regulatory policies on GM foods. Thus, while the United States has embraced biotechnology with open arms, creating a very permissive regulatory system, Europe has started to adopt an alternative regulatory pattern. Unlike the US model, the European regulatory model assumes a distinction between GM foods and conventional products, labeling, expanded risk management, and segregation and coexistence policies. These two regions have created such different regulatory policies for GM foods that many take them as opposing models for regulating these products. For authors like Bernauer (2003), the two regions have created a regulatory polarization in the world today. Among the different factors that are to influence the development of these systems are the very political disputes that have occurred around GM foods in these two regions of world geopolitics and how political systems have ultimately responded to them. Thus, the controversies associated with GM foods have produced distinct regulatory models. In the final Chap. 7 the existing regulatory policy for GMOs in Brazil will be examined taking into perspective the existing regulatory disputes between the United States and Europe. These two models (United States–Europe) will be used to better understand the type of regulation that exists today in Brazil.

---

[14] For an analysis of these different meanings, see Baldwin, Cave and Lodge , (2012).

3.7 The Politicization of Science

The European moratorium involved a reconfiguration of the role of science in the regulatory policy framework for GM foods. This seems to show that disputes about the risks of these products tend to have repercussions on the assessment of the use of scientific knowledge in decision-making about biotechnology. This is because regulatory policy seeks to use a scientific approach to measuring the risks associated with GMOs. This approach originates in fields such as engineering, economics, and epidemiology that generally seek to provide an objective reading of risks (Thayill, 2014, p. 64). These models seek to find the causes of existing risks and establish predictions for the undesired effects. In this approach, risk is defined as the "probability that a particular adverse event (identified as a hazard) occurs during a stated period of time" (Thayyil, 2014, p. 65). This approach involves different steps such as hazard identification and characterization and exposure assessment. The model also implies a bifurcation in risk regulation in terms of decision-making. From this model, a distinction is made between risk assessment and policy management. The techno-scientific approach to risk, and the bifurcation it implies, suggests a somewhat simplified division between fact and value. It presumes that RA offers objective scientific data. And these can guide policy decisions almost directly. However, as some works point out, risk assessment itself contains normative elements which, in a way, should make us suspect that these analyses present themselves as a kind of scientific information in the usual terms (Cranor, 1997). This, in turn, should call into question the possibility of separating risk (science) and management (politics) in a simplistic way. This is because the RA themselves can be contaminated with value judgments.[15]

These questions about the relationship between science, ethics, and politics reverberate in the controversies over GM foods. This is when the regulatory process is criticized for its technocratism in obscuring these issues under the guise of scientific RA. In this framework, the very role of science in the regulatory process is called into question. Hence, science that provides informative guidance to the regulatory process should not be recognized as science in strict terms. It should be seen, perhaps, as a type of normative science.[16] It has become evident to many

---

[15] Some government reports that seek to offer guidelines for dealing with technological risks are currently contradictory on some of these issues. As Thayyil shows, based on the analysis of Adams (1995) the distinction between fact and value is reiterated in some of these documents while acknowledging the problems existing in such division. See, on this point, Thayyil (2014) and Adams (1995).

[16] The vision of a "normative science" is not reducible to regulatory science as it seems to apply to academic science as it is known. In any case, the view of a "normative science" has a relationship to ethical and political issues in much the same way that regulatory science generally does. Verhoog (1993) sees normative science as follows: "We can speak about 'normative science' when scientists take serious notice of the normative aspects of sciences, both in theory and in practice. 'In theory' means that they must try to develop a logic or methodology of ethical reasoning. By 'in practice' is meant that discussions about these normative aspects should not be limited to the scientific community. The increase in normative rationality should go together with an increase in the 'legitimacy' of the decision-making process, with the attempt to make decision making about normative issues more democratic" (Verhoog, 1993, p. 96). In a way, we could say that the scientific knowledge that is usually used in regulatory policies is even more involved with normative and political issues than the academic sciences. Therefore, it makes sense to consider that regulatory science should be examined and evaluated from its connections to these same issues.

environmental policy analysts that the disinterested character of science can no longer be taken for granted. While this may be the case for science that is practiced in universities, the issue becomes even more sensitive for regulatory science. For this reason, a post-normal model of science has been advocated in some places in the regulation of GM foods. In the classical model of science, one pursues only one epistemic goal, while post-normal science deals with multi-epistemic goals. This makes it necessary for science in the regulatory process to develop in a multidisciplinary manner (Thayyil, 2014). Unlike classical science and the old models that relied on it to regulate technological innovation, there are no general expectations in this approach about "objective standards and absolute truths during regulation of risk" (Thayyil, 2014, p. 21). One place where a significant shift has been taking place on these issues is the Netherlands. For over two decades the country has been implementing a change in its Environmental Assessment Agency, transitioning from what Petersen et al. (2011) call a "technocratic model of science advising" to move towards a "post-normal science" (PNS) model.

3.8 The Labeling Conflict

According to McGarity (2007, p. 128), the labeling is one of the most controversial issues fueling the public debate about GM foods. Many might consider labeling to be a residual issue in this debate, but it is easy to see why it has become so important in the controversies over GM foods. Many of the ethical, scientific, and regulatory issues that have been examined above tend to be expressed in the labeling policy. It is through labeling that companies and governments establish a relationship with the consumer. Whatever the answer to the ethical questions that have been seen above, part of it will involve some kind of policy for labeling GM foods. For example, if consumers are considered to have the ethical and legal right to make their own decisions about the consumption of GM foods—the issue of consumer freedom—then the answer to the question will come largely through labeling policy. For if consumers have a right to choose whether to consume GM products and if it is expected governments have a duty to provide them with this opportunity, then it will be recognized that these things can only be achieved by a policy of proper labeling of GM foods. It is through labeling, then, that consumers, in theory, will be able to make their consumer choices. On the other hand, if it is considered that the government should establish some type of policy to monitor and supervise these products within the consumption network for reasons of food and environmental security, then this will only be possible through an appropriate labeling system for GM foods. Therefore, as Weirich (2007) points out, controversies over the marketing of GM foods raise questions about the very objectives of labeling policy. The questions associated with the debate about the ethics and risks of GM foods trickle down to our views about the meaning and function of the labels that are used to classify foods. What foods should be labelled? And how should this labeling be implemented? It is not clear whether labeling should be limited to providing information on the nutritional safety of the food, as is the case in countries like the United States, or whether it should provide a broader range of information that goes beyond a mere concern with the nutritional basis of the products that will be sold to consumers. Or whether the labeling system can incorporate safety concerns that transcend the scope of conventional systems. Thus, as Weirich (2007) will note, depending on how labeling may be defined and implemented, environmentalists may avoid GM foods not only for reasons associated with nutritional issues, but because of the implications this innovation may have for traditional lifestyles and also for environmental reasons.

Labeling, like the other issues examined in this chapter, is another dimension that, when considered, may raise a number of contentious issues. In a market economy, accurate consumer information tends to be a critical aspect of ensuring the necessary economic confidence in the market (McGarity, 2007, p. 128). At the same time, the topic raises the issue of government regulation and intervention, since in many countries the regulatory regime allows government agencies to protect the consumer from fraud and misleading advertising. In the case of the controversies raised by GM foods, two issues, as McGarity (2007) points out, tend to prevail. First, there are questions about the power of the law to require companies to label foods with information pointing to the presence of GM materials. Second, there is the issue of the reliability of the information that is passed on to consumers.

Having examined some of the dimensions usually present in disputes over GMOs, the final part of this chapter offers an introduction to the conflict involving the commercial release of RR soybean in Brazil. In this introduction, an attempt is made to provide a timeline of some events that help locate the conflicts that will be examined in subsequent chapters of the book. Thus, some dates and events are presented to help compose a narrative of the early phase of the conflict. The details and developments of these disputes will, as indicated, be taken up in the next three chapters.

3.9 The Commercial Release of RR Soybean in Brazil

One of the first records involving the planting of GM seeds in the world occurred in 1994. But it was only two years later, in 1996, that a larger and more significant area was planted, covering 1.66 million hectares in the United States. Some reviews consider it to be the world's first commercial GM crop (Brookes, 2014, p. 25). Since then, GM crops have only increased on a global scale and by 2010 had already reached the level of 139 million hectares on the planet. This corresponds to 71% of the area used for agriculture in Europe (Brookes, 2014).

RR soybean, whose commercial release in Brazil will be examined in this book, was one of the first GM seed varieties widely adopted on a global scale. Almost the entire area of GM seed crops in the world today is now confined to soybean, maize, cotton, and canola. Taking the data from the first decade of the twenty-first century, one has the following information. Of this total, soybean had 51% of the planted area, followed by corn (30%), cotton (14%), and canola (5%). Among other types of plantations made with GM seeds, soybean is the one with the largest planted area in the world today. In 2002, 50% of world production used GM soybeans. By 2010,

3.9 The Commercial Release of RR Soybean in Brazil

this percentage had risen to 70%. In most countries that use GM soybean, it represents 80% of production. In countries like Argentina and Paraguay, this proportion can reach 99% (Brookes, 2014, p. 25).

One of the factors that ended up favoring the conflict over the release of GM soybean in Brazil is linked to the economic role played by the country in the global agricultural market. Brazil is the third largest agricultural exporter in the world, behind only the United States and France. And in the international soybean market, Brazil is the second largest exporter. So, as Paarlberg will say: "All of Brazil's policies regarding GM crops and foods are significantly conditioned by the importance of farm exports of Brazil's economy" (2001, p. 79). Thus, the policies created for GMOs in agriculture were designed to increase the competitiveness of agricultural exports in the country. In parallel with the country's economic power, expressed through its strategic interests in the agricultural system, a factor that will favor the conflict is society's growing participation and engagement with environmental issues.[17]

The conflict in Brazil will occur because of the commercial release of Roundup Ready (RR) soybean that had already been approved for planting in the United States. RR soybean is tolerant to the herbicide glyphosate whose commercial name is Roundup Ready.[18] RR soybean is a GM seed that has a glyphosate herbicide tolerance event. This soybean presents a trait and an event that suggests its differentiated condition as a GMO. The trait refers to the desirable attribute that it incorporates. In this case, the desirable trait is resistance to glyphosate herbicide. And the event is related to the genetic manipulation operated on the seed itself that allows the trait in question to be promoted.

The controversy over the commercial liberation of transgenic soybean (Tg soybean) would erupt at the end of the 1990s. The conflict, in the Brazilian case, occurred, with a slight delay, when the case is compared, for example, with Europe and Asia. For, as Leite observes, when the controversy broke out in Brazil, in "Europe and Asia, transgenics literally caught fire, with environmental activists setting fire to fields cultivated with varieties of genetically modified organisms" (2000,

---

[17] As Paarlbert (2001) will note, movements associated with environmental protection and consumer rights in the country began to interact with political party leadership, as well as using the media and the judicial system itself in order to make their voices heard. These movements, and the political articulation that they developed in the period, expressed the same capacity for resistance and organization as the political movements that developed in Europe and elsewhere in the world.

[18] When glyphosate first emerged, the herbicide was most often used in periods before planting, or, in many cases, applied with specific equipment and techniques to avoid contact with the seeds. However, between 1992 and 2002, its use increased more than sixfold, making it the most widely used herbicide in the United States. The increase is associated with the increasing use of seedlings that are tolerant to the herbicide (Cerdeira and Duke, 2006). Some work indicates that glyphosate use increased by approximately 7% in the United States after the introduction of RR seeds. In the paper presenting this data, the author points out that "Contrary to often-repeated claims that today's genetically-engineered crops have, and are reducing pesticide use, the spread of glyphosate-resistant weeds in herbicide-resistant weed management systems has brought about substantial increases in the number and volume of herbicides applied" (Benbrook, 2012, p. 01).

10). This was the first license granted by CTNBio for the commercial-scale cultivation of a GM seed, although the commission had released the use of other GM seeds for other purposes (experiments) with more stringent safety conditions.[19] The commercial release of RR soybean can be considered the fuse of the conflict over GM foods in Brazil, as a set of events, legal disputes, and conflicts will develop from this decision. The conflict will be marked by a judicialization of the issue, as CTNBio's decisions will be challenged in the courts. On the other hand, in this process, legal injunctions will arise to support and sustain the CTNbio decision.

The conflict will be projected at different dates and in different regions of the country. And, as will become evident in the next three chapters, it incorporated different issues that touched on the dimensions examined in this chapter. In some areas, the conflict was much more dramatic than in others, as is the case in the south of the country, where, as Leite (2000) observes, it almost took the form of a regional war. The conflict entered the media (Internet, TV, and radio), the congress, TV programs, and legal system through legal disputes. Getting a clear picture of all these events is somewhat difficult, but one can map the initial phase from some important dates and events.[20]

CTNbio was created in 1996, but the conflict over the release of GM foods that is examined in this book began in 1998. Even before CTNBio's decision on RR soybean took place, there were signs of the conflict that would soon erupt in the country. In February 1998, for example, the Federal Police, through an anonymous tip, found GM soybean seeds at the Passo Fundo (RS) airport. During the period, those responsible for the importation were not identified, and suspicions arose that the GM soybean in question had been imported from Argentina, because the country was already ahead of Brazil in the planting of GM seeds. Since at the time there was no license for the commercial use of this type of product, the use of the soybeans was considered illegal, although the owners of the seeds were not identified at the time.

In June 1998, Monsanto submitted a request to CTNBio to allow the commercial cultivation of RR soybeans. But at the same time, other similar processes were already underway.[21] The RR soybean application will be the first to be filed for large-scale commercial use. In September 1998, more precisely on August 16, the 11th Federal Court of Appeals, appealing to the precautionary principle, granted an injunction to the Consumer Defense Institute (Idec), snapping the ban on commercial planting of soybean. A few days later, on September 16, CTNBio will make a decision pointing out its disagreement with the legal decision. The decision intensified the existing conflict because, in a way, the commission was issuing the opposite

---

[19] The CTNBio (National Technical Commission on Biosafety), an agency linked to the MS&T, was created to examine the safety of genetically modified organisms (GMOs). The organization is responsible for issuing opinions and licenses for the commercial use of GMOs in Brazil.

[20] For a chronology of events associated with the conflict, see Menasche (2002).

[21] In July 1998, more specifically on July 24, the 6th Federal Court of Brasília granted a preliminary injunction filed by Greenpeace, which claimed the suspension of the commercialization of oil made from transgenic soybeans. See Menasche (2002).

3.9 The Commercial Release of RR Soybean in Brazil

judgment from the one issued by the 11th Court of the Federal Court, since, in its decision, CTNBio indicated that all the necessary requirements by the national biosafety law had been applied in the decision to release RR soybean. Thus, the commission indicated that there were no substantial risks that prevented the release of the product. In the decision, 13 of the 15 CTNBio members present voted for the release. Of this group, one member representing consumers voted against the release while the representative of the Ministry of Foreign Affairs abstained in the decision. This then validated the release of RR soybean for commercial use. The judge's decision, in turn, indicated that these requirements were not being met in the process.

In view of the existing disagreements that crystallized between CTNBio and the courts, public hearings were held still in 1998 to debate the commercial release of GM foods in the country. The public hearings were promoted by the Consumer Defense, Environment, and Minorities Commission of the House of Representatives. The debates will be marked by conflicting views on transgenics, although a favorable view on certain points, in particular the need for labeling of GM foods, will predominate in the discussions. Two days after these public hearings, the previous injunction that prevented the commercial authorization for RR soybean will be overturned. This judicial injunction not only overturned the previous prohibition but, at the same time, ruled out, in line with CTNBio's decision, the need to conduct environmental impact assessment and report (EIA-EIAR). A requirement that was presented in society's demands (Idec, Greenpeace) as well as in the injunction that, at first, prevented the release of RR soybean. In this period, the governor of Rio Grande do Sul, Olívio Dutra, will initiate his campaign against transgenics. Still in 1998, the legal conflict suffers a new setback. This time it was in favor of the groups that resisted the commercial release of GM soybean. In December 1998, Idec and Greenpeace obtained a preliminary injunction from the 6th Federal Court of Brasília which prevented the commercial release of RR soybean once again. The injunction made the segregation of GM crops and the labeling of GM foods compulsory.

During this period, in December 1998, the conflict in the south of the country was taking shape beyond the legal issue. This is because the government started to act proactively to prevent the planting of GM soybean in the region. It will do this through legal means, seeking to create specific legislation for the region, but at the same time seeking to legally coerce large farmers who were in favor of planting RR soybean. This will be done with threats that indicated his intention to supervise the planting of RR soybean in the region. A detailed analysis of this conflict is presented in Chap. 4. This chapter examines the beginning of the conflict that took place in the south of the country and the developments that ended up taking place. In this conflict, the government and civil society organizations were against the release of RR soybean, and large farmers in the region were in favor.

In June 1999, the ministry of justice will present a project for the labeling of GM foods in the country. The document covered only packaged foods and foods intended for final consumers. It did not include raw foods, foods without food additives, and industrially processed foods. So, as Paarlberg notes, the draft presented by the government did not introduce any new labeling for GM foods. It required only some subtle changes to existing labels. The government's labeling proposal was carefully worded so as not to require the segregation of GM foods (Paarlberg, 2001).

The first semester of 1999 will be marked by the attempt to reorganize several political groups. In April, the government will announce a meeting with the government ministers who were part of the CTNBio. The meeting was held with the objective of building a more unified discourse on the issue of GM foods on the part of the government. In the following month, 27 state secretaries met in Recife (capital of the state of Pernambuco) to discuss the issue. At the end of the meeting, the secretaries presented a motion indicating the group's disagreement with the release of GM foods, emphasizing the need for safety measures for their release.

Before concluding this chapter and moving on to the existing analyses in the following chapters, a brief remark will be made about the terminology to be used in the following parts of the document. While this chapter has made frequent use of the term "GMO" and "GM foods" to address the controversies developing over modern biotechnology, in analyzing the Brazilian case, the word "transgenics" will tend to appear in the speech of those who directly participated in the conflict in Brazil. To the extent that the conflict in Brazil was triggered by RR soybean, which can be defined as a transgenic, the very debate and existing analyses of the issue in Brazil are marked by the significant use of this expression. Hence, that word (and not GMO) was chosen to compose the title of this book. The difference between GMOs and transgenic is not very significant, but it exists. A transgenic is an organism that has its genetic information modified due to the insertion of a gene from another organism that is not of the same species. For example, in the case of RR soybean, it is a seed that has received DNA fragments from another organism (Agrobacterium). Not all GMOs, however, are transgenic, as genetic modification does not always involve importing DNA fragments from a different organism. It is possible to alter the genetic information of a food by changing the information that already exists in it, without having to import a gene from a different species. Therefore, as is often stated in the literature, every transgenic is a GMO, but not every GMO is a transgenic.

3.10 Final Considerations

The initial conflict over RR soybean in Brazil will be marked by these events. This "hailstorm of appeals and preliminary injunctions," as Leite (2000) puts it, to characterize the beginning of the conflict in the country, will set the pace of the conflict in subsequent years. In the analysis that will be presented below, one seeks to examine the events that occurred precisely at this early stage of the conflict to examine the structural axes of this dispute. After all, why did RR soybean become so controversial in Brazil? What would be the disagreements and interests that started to divide Brazilians (politicians, farmers, etc.) on the issue? In the next three chapters, an answer to these questions is offered. In them, an attempt is made to show that the conflict over the release of RR soybean can be understood through three structuring axes. Among these structuring axes are issues associated with the (a) distributional effects (justice) of GM seeds in the agricultural system. They also relate to (b) scientific risk assessment and, finally, (c) the issue of labeling to be applied to GM foods. The issue of RA and its implications has gained great prominence in court cases and in the discourse of civil society organizations. The distributive issue is essential to understanding the conflict that has emerged in the south of the country. And the issue of labeling is an important one for understanding the influence that civil society ended up exerting on the regulatory process of GMOs in Brazil. So, in the analysis that follows, the study focuses on trying to present the structuring of these discourses on GM foods in the conflict. That is, it seeks to examine the story line of each type of conflict. The emphasis of the analysis will be on the dynamics of the political debate. In the following chapters, it will be examined how and why the actors who actively participated in these conflicts defended their positions. And, where possible, the implications of these positions for the course of the conflict itself will be considered. Although the analysis focuses on this early phase of the conflict, in some cases information linked to these conflicts in subsequent years is examined. But this information is integrated only to the extent that it helps us understand the story line of the structuring axes of the conflict that unfolded at the beginning of the dispute.

References

Adams, J. (1995). *Risk*. Routledge.
Altieri, M. A., & Rosset, P. (2002). The reasons why biotechnology will not ensure food security, protect the environment, or reduce poverty in the developing world. In R. Sherlock & J. D. Morrey (Eds.), *Ethical issues in biotechnology* (pp. 175–182). Rowman & Littlefield Publishers.
Amossy, R. (2021). *In defense of polemics*. Springer.
Baldwin, R., Cave, M., & Lodge, M. (2012). *Understanding regulation: Theory, strategy, and practice*. Oxford University Press.
Benbrook, C. M. (2012). Impacts of genetically engineered crops on pesticide use in the U.S – The first sixteen years. *Environmental Science Europe, 24*, 24.
Bernauer, T. (2003). *Genes, trade and regulation. The seeds of conflict in food biotechnology*. Princeton University Press.
Bodiguel, L., & Cardwell, M. (2010). Genetically modified organism and the public: Participation, preferences, and protest. In L. Bodiguel & M. Cardwell (Eds.), *The regulation of genetically modified organisms* (pp. 12–36). Oxford University Press.
Bridgstock, M. (1998). Controversies regarding science and technology. In M. Bridgstock, D. Burch, J. Forge, J. Laurent, & I. Lowe (Eds.), *Technology and society: An introduction* (pp. 83–107). Cambridge University Press.
Brookes, G. (2014). Global adoption of GM crops, 1995-2010. In S. J. Smyth, P. W. B. Phillips, & D. Castle (Eds.), *Handbook on agriculture, biotechnology and development*. Edward Elgar.
Bud, R. (1993). *The uses of life. A history of biotechnology*. Cambridge University Press.
Cerdeira, A. L., & Duke, S. O. (2006, September). Current status and environmental impacts of glyphosate-resistant crops: A review. *Journal of Environmental Quality, 35*(5), 1633–1658.
Comstock, G. (2002). Ethics and genetically modified foods. In M. Ruse & D. Castle (Eds.), *Genetically modified foods* (Genetically modified foods) (pp. 88–107). Prometheus Books.

Cranor, C. F. (1997). The normative nature of risk assessment: Features and possibilities. *Health, Safety and Environment, 8*(2), 123–136.

Dusek, V. (2006). *Philosophy of technology. An introduction.* Blackwell Publishing.

Giddens, A. (1994). *Beyond left and right. The future of radical politics.* Polity Press.

Horlick-Jones, T., et al. (2007). *The GM debate. Risk, politics and public engagement.* Routledge.

James, H. S. J., Jr. (2018a). Introduction. Ethical tensions and new technology: An overview in the context of agricultural biotechnology. In H. S. James Jr. (Ed.), *Ethical tensions from new technology. The case of agricultural biotechnology* (pp. 1–11). CABI.

James, H. S. J., Jr. (2018b). Ethical tensions from new technology. In H. S. James Jr. (Ed.), *The case of agricultural biotechnology.* CABI.

Jasanoff, S., & Kim, S.-H. (2015). *Dreamscapes of modernity. Sociotechnical imaginaries and the fabrication of power.* The University of Chicago Press.

Kenneth, D., & Thompson, P. B.. (2008). *What can nanotechnology learn from biotechnology? Social and ethical lessons for nanoscience from the debate over agrifood biotechnology and GMOs.*

Kim, T. (2014). Biotechnology: Regulatory issues. In N. K. van Alfen (Ed.), *Encyclopedia of agriculture and food systems* (Vol. 1, pp. 153–172). Elsevier.

Krimsky, S., & Gruber, J. (2014). Introduction. The science and regulation behind the GMO deception. In S. Krimsky & J. Gruber (Eds.), *The GMO deception. What you need to know about the food, corporations, and government agencies putting our families and our environment at risk.* Skyhorse Publishing.

Lee, K. (2009). Biology and technology. In J. K. B. Olsen, S. A. Pedersen, & V. F. Hendricks (Eds.), *A companion to the philosophy of technology.* Blackwell Publishing.

Leiss, W., & Chociolko, C. (1994). *Risk and responsibility.* McGill-Queen's University Press.

Leite, M. (2000). *Os alimentos transgênicos.* Publifolha.

Levidow, L. (2014a). European Union policy conflicts over agbiotech: Ecological modernization perspectives and critiques. In S. J. Smyth, P. W. B. Phillips, & D. Castle (Eds.), *Handbook on agriculture, biotechnology and development* (pp. 153–165). Edward Elgar.

Levidow, L. (2014b). EU regulatory conflicts over GM food. In D. M. Kaplan & P. B. Thompson (Eds.), *Encyclopedia of Food and agricultural ethics* (pp. 834–841). Springer.

Markert, L. R., & Backer, P. R. (2003). *Contemporary technology. Innovations, issues, and perspectives.* The Goodheart-Willcox Company, Inc.

McGarity, T. O. (2007). Frakenfood free: Consumer sovereignty, federal regulation, and industry control in marketing and choosing. Food in the United States. In P. Weirich (Ed.), *Labeling genetically modified food. The philosophical and legal debate.* Oxford University Press.

McGarvin, M. (2001, May). Science, precaution, facts and values. In O'riordan, T et al. (Eds.), *Reinterpreting the precautionary principle.* Cameron.

McHughen, A. (2008). Learning from mistakes: Missteps in public acceptance issues with GMOs. In K. David & P. B. Thompson (Eds.), *What can nanotechnology learn from biotechnology? Social and ethical lessons for nanoscience from the debate over agrifood biotechnology and GMOs* (pp. 33–54). EUA: Elsevier Inc.

Menasche, R. (2002). Legalidade, legitimidade e lavouras transgênicas clandestinas. In Alimonda, H. (org.), *Ecologia política: Naturaleza, sociedad y utopia.* CLACSO.

National Academics of Sciences, Engineering, and Medicine. (2017). *Communicating science effectively: A research agenda.* Tha National Academies Press. https://doi.org/10.17226/23674

Paarlberg, R. L. (2001). *The politics of precaution. Genetically modified crops in developing countries.* The Johns Hopkins University Press.

Petersen, A., et al. (2011). Post-normal science in practice at the Netherlands environmental assessment agency. *Science, Technology, & Human Values, 36*(3), 362–388.

Priest, S. H. (2005). Biotechnology. In *Science, technology, and society. An encyclopedia* (pp. 34–40). Oxford University Press.

Raynaud, D. (2015). *Scientific controversies. A socio-historical perspective on the advancement of science.* Transaction Publishers. REDE.

References

Reiss, M. (2001). Biotechnology. In R. Chadwick (Ed.), *The concise encyclopedia of the ethics of new technologies* (pp. 13–42). Academic Press.

Robinson, J. (1999). Ethics and transgenic crops: A review. *Electronic Journal of Biotechnology, 2*(2) Issue of August 15, 71–81.

Thayyil, N. (2014). *Biotechnology regulation and GMOs. Law, technology and public contestations in Europe.* Edward Elgar.

Thieman, W. J., & Palladino, M. A. (2020). *Introduction to biotechnology.* Pearson Education.

Thompson, P. B. (1997). *Food biotechnology in ethical perspective.* Champman & Hall.

Thompson, P. (2002). Bioethics issues in a biobased economy. In M. Ruse & D. Castle (Eds.), *Genetically modified foods* (Genetically modified foods) (pp. 68–76). Prometheus Books.

Timmermann, C. (2020). *Social justice and agricultural innovation.* Springer Cham.

UNECE. (1998). *Convention on access to information, public participation in decision-making and access to justice in environmental matters.* Endereço: https://unece.org/fileadmin/DAM/env/pp/documents/cep43e.pdf. Acesso em 20 Oct 2021.

United Nations. (1992). *Convention on biological diversity.* Available at: https://www.cbd.int/contact/. Accessed April 10, 2021.

United Nations. (2000). *Cartagena protocol on biosafety to the convention on biological diversity: Text and annexes.* Secretariat of the Convention on Biological Diversity. Endereço: https://bch.cbd.int/protocol/text/. Acesso em 20 Jun 2022

Verhoog, H. (1993). Biotechnology and ethics. In T. Brante, S. Fuller, & W. Lynch (Eds.), *Controversial science. From content to contention* (pp. 83–106). State University of New York Press.

Weirich, P. (2007). Introduction. In P. Weirich (Ed.), *Labeling genetically modified food. The philosophical and legal debate.* Oxford University Press.

Chapter 4
A Territory Free of Transgenics: The Conflict over the Release of RR Soybean in Southern Brazil

> Deep down, the issue of transgenics is being used as a front so that the areas [...] the productive plantations in Rio Grande do Sul, are susceptible to expropriation for Agrarian Reform, for the MST, since the Secretary of Agriculture of our state is a representative of the MST. (Union representative).

> Our government proposal, amply debated during the electoral campaign, prioritizes family agriculture, thus we cannot agree with a technology that exactly excludes small family farmers. This is a very strong reason why we are against the production and commercialization of transgenics. (Jose H. Hoffmann, Secretary of Agriculture).

> So, on the one hand, you have these social movements that are concerned about this not only from an environmental or public health point of view, but with other risks, with these other aspects. Which is what? Which is the domination, which is the control of the companies, which is the technological dependence that this generates from the economic point of view and from the technical knowledge point of view. (Agronomist of the Central Settlement Cooperative).

At the end of the 1990s, a conflict arose in Rio Grande do Sul (RS) involving environmental NGOs, entities linked to small farmers, farmers, and the different levels of government (state versus federal). The conflict was centered on the release of Roundup Ready (RR) soybean and, more generally, the commercial use of GMOs in the agricultural system. On one side of the conflict were farmers favorable to the use and commercialization of RR soybean and, on the other, organizations, parties, and the state government itself that sought to bar its planting and commercialization. After the resistance sponsored by the RS government, similar movements also took place in other parts of the country. As a result, in 2001, restrictions on the GM crops emerged in Paraíba, Bahia, Minas Gerais, São Paulo, and Espírito Santo.

The conflict reached its peak in 1999, when Olívio Dutra's Workers' Party (PT) government launched a campaign whose main objective was to transform the state of Rio Grande do Sul (RS) into a "Um Território Livre de Transgênicos".[1] The goal of the campaign was to ban the commercialization of GM seeds throughout the state's territory. Judging the campaign by its stated goal, it could be said that it resulted in a major failure. From "Um Território Livre de Transgênicos," the state ended up becoming a region free for the commercialization of GM feeds. Currently, RS is one of the largest producers of GM soybean in the country. Judging the campaign from a political perspective, on the other hand, one can point out its important effects. The state became a symbol of resistance to GMOs and helped broaden the debate on the social impacts of these products on a national and international scale.

This chapter seeks to show that, more than issues involving the risks and environmental safety of RR soybeans, the conflict that occurred in RS was largely associated with distributive issues involving land reform. The conflict expresses several characteristics that allow this argument to be sustained. In the period, the RR soybean was interpreted as a threat to the agrarian reform that had become Olívio Dutra's (PT) political banner in his electoral campaign for the state government. It is for this reason that this conflict does not refer us to an issue merely of the environmental safety of GMOs, but suggests the radicalization of existing agricultural conflicts in the region that predate the very issue of the commercial release of RR soybean. How these preexisting agricultural conflicts came to be intertwined with the issue of GMOs is what will be examined in this chapter. The favorable position for the commercial release of RR soybeans will be denominated in this part of the work as alliance or agricultural modernization discourse. The group of actors who sought to paralyze this release will be referred to here as alliance or discourse of EJ.

But before examining the conflict that occurred in southern Brazil, a preliminary analysis of the issues that make up the concept of EJ, GMO, and agriculture will be offered. The analysis of this conflict will offer an introduction to how issues that have been linked to the conflict can be understood. It will be examined below, therefore, three important points involving this conflict. First, the concept of EJ and the relationship between environmental risks and distribution issues will be examined. And from this first part, some considerations about the distributive impacts of GMOs in agriculture and how this issue was projected in the conflict that occurred in the South in the period 1998–1999 will be presented (Table 4.1).

---

[1] In English it can be translated as "A GMO-Free Territory" or more specifically, "A Transgenics-Free Territory." During the period of conflict, the expression "transgenics" was incorporated into the campaign slogan of the Olívio Dutra government against the use of GMOs in agriculture. The term was used more than "GMOs," and, in many cases, although there are differences in meaning between these two expressions, transgenics and GMOs were taken as synonyms.

4.1 The Concept of Environmental Justice

Table 4.1 The transgenic conflict in the south of the country

Structuring issue	Alliance of agricultural modernization	Alliance of environmental justice
RR soybean risk	"Positive-sum game." Nonexistent trade-off thesis.	"Negative-sum game." Emphasis on the economic and cultural threat on small farmers.
Origin of risk	The risks of RR soybean are equivalent to the risks of conventional soybeans.	Cultivar law, patent law, and biological composition of soybeans that prevents the reuse of seeds by small family farmers.
Consequences of RR soybean	Agricultural modernization (higher productivity).	Concentration of ownership (land, seeds) to large corporations.
Biosafety law	Favors the commercial use of RR soybeans with RA.	Requires EIA for commercial release of RR soybean.
Sustainability	Safe use of GM seed in the agricultural system as sustainable.	Family farming and agroecology as axes of sustainable agriculture.
Agricultural systems	Preference for large-scale export agriculture. No concern for family farming	Modern large-scale agriculture and family farming as incompatible. Emphasis on alternative agricultural models (e.g., agroecology).
Rationalization	RR soybean as a positive result of the agricultural modernization process (rationalization of agriculture).	RR soybean as a dangerous outcome of the "green revolution" (resistance to rationalization).
Marketplace	Focus on the GM seed market (EUA)	Focus on alternative agricultural market (no GM seeds—Europe)
Agrarian reform	GMOs (RR soybean) do not pose a risk to the agrarian reform.	GMOs (RR soybean) as a threat to land reform.
Risk response	Risk analysis and RR soybean as safe.	Agroecological family farming as a fair and secure agricultural system.
Science	Scientific knowledge confirms safety of RR soybean.	Scientific doubts about the safety of RR soybean.
Regulation of GMOs	Competence of the federal government in regulating GM seeds (RR soybean).	Regional autonomy in the regulation of GM seeds (RR soybean) in the formulation of agricultural policy.

Source: The Author

4.1 The Concept of Environmental Justice

The EJ movement can be seen as one of the most important wings of contemporary environmentalism. Its origin is associated with the struggles that communities of US workers began to undertake against toxic pollution since the second half of the twentieth century. A second interpretation of the movement also establishes its origin in the mobilizations of black communities in the United States that began to fight against the unequal distribution of environmental risks[2]. This view of EJ which links it to distributional conflicts associated with environmental risks, reflects the most common point of view that can be found in the literature on the concept. It is expressed in Shrader-Frechette's conception, who states that distributive justice is essential to the pursuit of EJ:

> because it requires a fair or equitable distribution of society's technological and environmental risks and impacts. It refers to the morally proper apportionment of benefits and burdens—such as wealth, opportunity, education, toxic waste dumps, dirty air, and so on—among society's members. (Shrader-Frechette, 2002, p. 24)

So EJ has been seen as involving both equitable distribution of environmental goods and evils and also greater public participation so as to provide conditions for political decisions of this kind. From this perspective, EJ basically involves two dimensions: a distributive dimension ([a] how are environmental benefits and costs distributed?) and another participatory ([b] how are these distributive decisions made?) (Figueroa & Mills, 2003).

For approaches that fall within the EJ perspective, environmental inequalities are seen as associated with economic inequality (class) and cultural factors (e.g., environmental racism). In both these cases, the inequality in question is seen as a reflection of an unequal distribution of environmental risks whose causes lie in distinct social factors.[3] For the first view, which sees environmental inequality as associated with social classes, the cause of such conflicts would be in the unequal distribution of wealth. The first is a direct reflection of the second. In the second case, environmental inequality would have its origin in differences of cultural status existing in society and in discriminatory and racist practices that arise from these differences. It is argued here that environmental risks are disproportionately distributed among different groups, causing the burden of pollution to fall most heavily on black communities or other marginal social groups. In this case, the cause of the maldistribution of environmental risks would be founded in what is called environmental racism.[4] This also suggests that environmental inequality in this case is a direct reflection of cultural problems in society.

Beyond distributive issues, the idea of EJ has also incorporated a participatory dimension. This is because inequality has also been considered a result of power asymmetries existing in the political decision-making process. This sense of the

---

[2] Although some academic works establish a specific origin for the movement, some authors seek to point out the diversity of popular political movements that integrate it internally. These include (a) the civil rights movement, (b) the health movement and occupational safety, (c) the indigenous land rights movement, and (d) the solidarity (rights and self-determination of people). About this diversity of the movement of EJ, see Schlosberg (2007).

[3] For an examination of EJ as a reflection of racial inequalities, see Bullard (2004). And for a examination of environmental inequalities as expressions of class differences, see Gould (2004).

[4] Robert Bullard, one of the leading environmental justice scholars in the United States, will define environmental racism as "environmental public policies, practices or directives that affect differently or harm (intentionally or unintentionally) individuals, groups, or communities of color or race" (Bullard, 2004, p. 42).

4.1 The Concept of Environmental Justice

idea of EJ directs its attention to the distribution of power of different social groups in the political decision-making process. The unequal distribution of environmental risks could thus be seen as resulting from an inequality that occurs in terms of the political power that is accessible to the different groups in the process. In this way, Shrader-Frechette (2002, p. 28) links the distributive dimension of the EJ concept to a participatory dimension, suggesting that the latter is associated with procedural norms that offer guarantees to people the same opportunities in the decision-making process.

However, there are several important points to consider when examining these inequalities. If distributional inequality is linked to inequalities of political power in decision-making spheres, this same inequality may be linked to differences in the recognition that these groups enjoy in social life. As Schlosberg (2007, p. 65) will note: "a lack of individual or cultural recognition and a lack of valid participation in the political process". Which means that one can then, in certain situations, look at the lack of social recognition as an important factor in the distribution of environmental risks (Schlosberg, 2007, p. 59). Recognition has therefore been seen as a defining element of the concept of EJ. although the topic receives little attention in international and national work.[5] The concern with recognition has arisen especially in work that seeks to extend how the identity of social groups influences distributional conflicts in this area. For these works, environmental inequality and the conflict it engenders tend to be the result of the absence of cultural and political recognition.

Finally, EJ can be interpreted from a particular view of the capacities of individuals and communities in their interaction with the natural and physical environment. It finds expression in Bryant's definition when this author defines EJ as being the "where people can interact with confidence that their environment is safe, nurturing, and productive. Environmental justice is served when people can realize their highest potential" (Bryant, 1991, p. 6). In this approach, concerns are not only directed to distributional issues, but to the processes that affect well-being and the capabilities and needs of individuals and communities. This perspective is very close to the approach offered by the concept of sustainable development in the Brundtland Report. In this work, sustainable development was defined as a process of satisfying human needs, which in turn were associated with the very idea of justice.[6] Having

---

[5] The low importance attached to the concept of recognition is not only related to the literature on EJ, but it is a problem that affects, to some extent, theories of justice in a way general. On the work of John Rawls, a prominent theorist on this topic, Schlosberg (2007, p. 21) notes that in his work "respect and dignity are preconditions for distributive justice," without the received substantive theoretical analysis. Consideration that he extends to the work of other authors and philosophers and for the literature itself linked to the concept of EJ.

[6] As Langhelle (1999, p. 140) writes: "Social justice can be seen as equivalent to the satisfaction of human needs, which, in turn, is what constitutes the primary goal of development, in sustainable development." Physical integrity and personal autonomy, two central needs which are present in the human needs theory of Doyal and Gough (1991), are found also in Nussbaum's capability theory. On the relationship between sustainable development and human needs, see Lenzi (2022), Haland (1999), and Langhelle (1999).

examined some of the points associated with the concept of environmental justice, I will next examine how some of these issues are linked to the use of GMOs in agriculture.

4.2 Distributive Impacts of Agricultural Biotechnology

Distributional issues are generally seen as an important point involving conflicts over the release of GMOs in agriculture. This is because their release can involve various distributional effects on the agricultural system, triggering subsequent social and environmental effects for different social groups (producers, farmers, consumers, and governments). In the context of the release of GMOs in agriculture, Thompson (1997) suggests the existence of the following axes involving the impact of agbiotech:

(a) The impact of agri-tech on small family farms.
(b) Impact on third-world countries.

The effects of agbiotech in poorer countries are generally discussed from three axes. It is believed that it may endanger the poorest farmers in the countryside through:

(a) Concentration of ownership ("fewer and larger mechanism").
(b) Global trade: increasing social and economic inequality between industrial agriculture and agriculture of the poorest producers.
(c) Intellectual property: creating inequalities through the mechanism of intellectual property.

While concerns about the concentrative tendency of large farms are an important element of the discussions in wealthy countries, this issue takes on a new dimension when raised in the context of countries where agricultural regions are characterized by extreme poverty. While the impacts on small farmers may involve issues of rights associated with the family farm structure, in poor countries these issues broaden out to raise concern over issues of malnutrition, food deprivation, and the diseases and ills associated with each of these. Let us examine some of these issues in the Brazilian context.

4.3 GMOs and Environmental Justice in Brazil

In Brazil, the emergence of the EJ movement is linked to two distinct moments. On the one hand, the movement is seen as being associated with actions undertaken in the past that had the distribution of environmental risks as the main focus of attention. Here, one could include, for example, the groups that fought for preventive

4.3 GMOs and Environmental Justice in Brazil

public health in Brazil in the past.[7] A second interpretation, however, sees the emergence of the movement in the political actions of different groups that began to incorporate more directly the precepts of EJ in the country. The emergence of the movement for this case would have several different points of origin. The collection entitled "Trade Unionism and Environmental Justice" published in 2000 by the Central Única dos Trabalhadores (CUT), together with IBASE, IPUUR, UFRJ, and the Heinrich Böll Foundation, could be seen as a starting point of the movement. It was then that, for the first time, the language of EJ started to be used in Brazil.[8] Subsequently, interest in the topic grew, leading, in 2005, to the holding of the International Seminar on Environmental Justice and Citizenship in Niterói (RJ), which brought together researchers, leaders of environmental organizations, and unions. The event demonstrated the global dimension of the EJ discourse, since the event served to disseminate the ideas and experiences of EJ that were linked to the American experience. From these experiences, the Brazilian Network of Environmental Justice (RBJA) was created, which has, among its principles, the following guidelines:

(a) Ensure that no social group, whether ethnic, racial, or class, bears a disproportionate share of the negative environmental consequences of economic operations, policy decisions, and federal, state, and local programs, as well as the absence or omission of such policies.
(b) Ensure fair and equitable access, direct and indirect, to the country's environmental resources (Brazilian Network of Environmental Justice apud Herculano, 2002).

Both in the United States and in Brazil, the actions of the EJ movement are perceived as a means of establishing a line of resistance to environmental inequalities. At the same time, this axis of the environmental movement began to incorporate the vision of environmental vulnerability from the conception of a rather generic subject, encompassing workers, blacks, the poor, small farmers, and indigenous people. Thus, while in the United States, EJ's discourse placed black communities as the main victims of the unequal distribution of pollution, in Brazil, this same discourse began to express a more plural vision of the process of victimization of environmental inequality.

In this framework involving the discussion of the idea of EJ in Brazil, there is little research in the country that associates this concept with small farmers' mobilizations. However, distributive issues involving the release of GMOs have been a focus of attention of the RBJA. The entity recognizes small farmers as a group that, among many others, is victimized by the process of agricultural modernization.

---

[7] The conference on EJ held in 2000 signaled that "a set of actions and social movements in the country can be identified as a search for 'environmental justice', even without using this expression" (Herculano, 2002, p. 6).

[8] It should be considered that, in this same period, several academic works already sought to operationalize the concept of EJ in order to apply it to the Brazilian context. On this point, see Acselrad (2010).

Acselrad (2010, p. 111), for instance, when discussing environmental inequalities in the countryside, argues that small farmers end up having their activities made unviable "due to the expansion of genetically modified soybean." Because of these inequalities, one of the objectives of the RBJA in the country is linked to the defense of "the rights of rural populations," while seeking an equitable environmental protection against "socio-territorial discrimination and environmental inequality" (RBJA, 2010, s/p). The movement also includes the defense of "valuing the different ways of living and producing in the territories, recognizing the contribution that indigenous groups, traditional communities, agro-extractivists and family farmers make to the conservation of ecosystems" (RBJA, 2010). In these and other passages, small farmers are presented as one of the main victims of the environmental inequalities that are expressed in the Brazilian agricultural reality.

In a document in which the principles of the movement in the country are presented, it is reported that:

> The traditional populations of extractivists and small producers, who live in the regions of the expansion frontier of capitalist activities, suffer the pressures of compulsory displacement from their areas of residence and work, losing access to land, forests and rivers, being expelled by large hydroelectric, road or mineral exploration, logging and farming projects. (Launching Declaration of the Brazilian Network of Environmental Justice) (Brazilian Network of Environmental Justice apud Herculano, 2002).

In general lines, the relationship between the GMOs and small farmers occurs tangentially in authors and documents that address the subject of EJ in the country. In any case, it is possible to find works that hint at a relationship between these issues, as these passages indicate. There are many studies carried out in the country that address the inequalities existing in the field. However, much of this discussion is not based on the concept of sustainability, linking the idea of EJ to this concept, and there are still few studies that integrate these terms with the issue of GMOs. For example, when Acselrad (2010) sees small farmers as victims of environmental inequalities due to the "expansion of transgenic soybean," the concern seems to be directed towards the unviability of "the activities of small organic farmers" (Acselrad, 2010, p. 111). In the RBJA, references to small farmers are common, including them as one group among many others that suffer from environmental inequality in the country. Having examined in this first part some of the issues surrounding GMOs and justice, the following will examine how these distributional issues came to influence the conflict over the release of RR soybean in the late 1990s.

4.4 Trade-Off: When Everybody Wins

For advocates of the commercial release of RR soybean, rather than being associated with the use of GMOs in agriculture, the greatest risks were seen to lie in the legal prohibition of their commercial use. In the late 1990s, when RR soybean was commercially released, the fears of the agricultural modernization alliance focused their concerns on the undermining of agricultural innovation in the country. And,

4.4 Trade-Off: When Everybody Wins

also, for the technical and scientific development of agricultural biotechnology itself. In the agricultural modernization alliance, economic losses were seen as integrated with scientific losses due to the common projects that were, and still are, in formation between biotechnology companies, universities, and government. In addition to being associated with a process of agricultural modernization that could offer benefits to the entire nation, RR soybean was also seen as bringing a kind of environmental modernization in agriculture. RR soybean was perceived as bringing lower environmental risks and simultaneously contributing to the promotion of sustainability in agriculture.

Launched by Bresser-Pereira (1999), the trade-off thesis allows us to understand the position that advocates of the release of GMOs have come to hold on the risks associated with RR soybean. The trade-off thesis indicates an attempt to examine the risks associated with the release of GMOs through a balance of costs and benefits brought about by the release of RR soybean. In May 1999, Bresser-Pereira (1999), in the midst of the conflict over commercial release, wrote a paper postulating that there were no social, economic, or environmental losses from the release of RR soybean. Thus, its practically nonexistent risks could not be evaluated in isolation, but based on a balance of gains and losses in which the former should be examined in light of the latter. However, since, in the view of the MS&T and CTNBio, the prevailing perception was of additional gains to the country's economic development, without the existence of visible losses, the release of RR soybean was seen as involving a "positive sum game." Bresser Pereira, in the period, will then state that:

> Regarding transgenic products, I believe that we should examine the problem from two points of view of the MS&T: one is specific to biosecurity and the other is the economic sphere, efficiency, technology and competitiveness of Brazil internationally. Both matters are obviously of interest to the Ministry of Science and Technology. We know that when we think about related objectives, there are complementary ones: on one side there is pure and simple economic development, and on the other, sustainable economic development. Two similar things, but different. And we sometimes have to make trade offs, bargains, to achieve development goals. The fight against pollution, for instance, is a costly program, involving immense technological difficulties, but it needs to be carried out. This is a situation that involves trade offs, because it is fundamental to defend the environment, nature, the world in which we live, and to think in the long term, for future generations. Now, in relation to transgenic products, as much as I analyze the problem and for everything that I have heard from all the technicians and scientists with whom I have talked, I do not believe that there is any trade off involved in this issue. There is no trade off in relation to economic development and self-sustainable development. There are no losses in terms of biological security and environmental protection, as a counterpart to gains in economic development. (Bresser-Pereira, 1999, s/p)

It is important to note that in the period of the RR soybean release, there was no intention by the MS&T, and much less by CTNBio itself, to examine closely any economic impact that the release of GMOs could have on weaker economic sectors. At least that is what can be gleaned from the minister's and CTNBio's discourse. For the view that is presented here is one where RR soybean is not seen as implying any kind of loss, either economic or environmental.

This view can be said to be the one that prevails in the policy of genetically modified organisms (GMOs) to date. With respect to small farmers in particular, there has been no concern to try to assess the impact of these products on their economic structure and farming practices. CTNBio produced no such assessment in the period of the RR soybean release because of the factors that influenced its own creation. Its main task would be to conduct an assessment of the environmental safety of GM seeds. These restrictions associated with its role have also influenced debates about the labeling of GM products, which, for some of the agency's past chairpersons, has been a "political" rather than a "technical" issue. The environmental RA that will guide the RR soybean release, for example, did not involve any consideration of its economic consequences on the Brazilian agricultural structure. The view that became predominant in CTNBio was that its responsibility would be limited to examining "scientific" issues involving the environmental safety of GMOs and not political factors such as labeling and its economic impacts on agriculture.

Since the release of RR soybean in the country, issues of economic and environmental justice have been seen as alien to the state regulatory bodies responsible for releasing GMOs. The RA, the main instrument for analyzing the environmental safety of GMOs in the country, for example, did not involve any such consideration. This stance of the government in the period, and which has remained unchanged over the last few years, is distinct from that which can be found in countries like Austria, where the economic impact on small family farms has become an important topic in the regulatory process of GMOs in that country, as it will also be one of the factors influencing the European moratorium on them. Austria, for example, bases its GMO regulation process on the principle of sozialverträglichkeit (social sustainability), which assumes that no agricultural innovation should impose a disproportionate burden on different groups in society. This principle informs that decisions on technological innovations in agriculture should consider, whether for social, moral, or economic reasons, the disproportionate burden that these same decisions can bring to different social groups in society (Seifert & Torgersen, 1997). In the country, since the commercial release of RR soybean, there has been no government decision that could even come close to this principle. It is necessary to consider that this principle does not even exist in Brazilian legislation and regulatory guidelines associated with agricultural biotechnology. The thesis that prevailed in the Brazilian government in the conflict was the existence of a "positive sum game" in the process of releasing these products, in which all players in the agricultural system would be immune to any type of loss. Including here even small farmers. Their impacts are strictly positive, bringing economic and even environmental benefits.

It is for this reason that the conflict that arose in the south of the country will have special importance for the debate on GMOs in the period. Although many interpretations of this conflict reduce it to a simple dispute involving defenders and supporters of GMOs and their possible environmental risks and the safety issues that can be detached from this question, it involved issues directly associated with agrarian reform in the country and thus brought to the fore the debate on distributive issues ignored in the process of regulating biotechnology. To understand the conflict that occurred in the south of the country, it will be necessary to return to the conflicts that unfolded there in the 1990s. Some of the actions and tensions that took place in the commercial release of RR soybean actually reflect a pattern in the shape of conflicts that precede this conflict. Thus, the meaning of some of these actions can only be understood by examining some of the facts that underpinned agrarian conflicts in the RG state even before the release of GM seeds was announced. These disputes are examined below.

4.5 GM Seeds and Land Reform

Olívio Dutra was elected representing the Popular Front: a network of organizations and parties that supported his campaign for governor of the state. The MST and other organizations such as Contag, AS-PTA, and Via Campesina were entities that supported his election. Several MST members at the time were affiliated to the governor's party (PT) and the government itself demonstrated an ideological and programmatic alignment with the discourse of these organizations, especially the MST. This alignment was so close that parliamentary advisors and key government positions were filled by people directly linked to the movement for the strengthening of family farming and the defense of alternative technical proposals involving agroecology.[9]

Until the first semester of 1998, GM seeds, and GMOs more generally, did not constitute an object of conflict in the state of RG. When examining newspaper news in this period, Heberle (2005, p. 188) notes that the "idea of transgenics […] did not yet register, in the state, great resistance." Even if it is not possible to observe any significant resistance against GMOs until this period, the news involving the use of biotechnological innovations in agriculture were becoming more and more frequent in the local newspapers. Overall, predominant in this period was news that merely signaled the agricultural productivity gains linked to agbiotech that were, on the whole, popularized by the local media.[10] At the same time, although it was possible to find news that signaled to the benefits of agbiotech innovations, there was still no concrete case that pointed to its commercial use in the state of RS. Moreover, the GM crops had been absent from the political agenda of the political parties until then. As Bauer (2006, p. 224) will note on the outbreak of the conflict in 1999: "An

---

[9] As Ros (2006, p. 238) will indicate: "Such attunement resulted in the occupation of strategic positions of the government by members of these movements, or by people linked to them, especially in the Secretariat of Agriculture and Supply."

[10] In examining media coverage of GMOs through 1997, Heberle (2005) notes that few are the reports that sought to point out the risks associated with these products. There are, apparently, two reasons for this. In general, press coverage generally referred to events that took place in the countryside outside the state of RS and were presented by the newspapers as scientific facts alien to the reality of local agricultural practices, a situation that will change with the news indicating the planting of GM seeds in the state and with the decision to commercially release RR soybean in the late 1990s.

issue that was not even on the electoral agenda" subsequently gained government policy status with the election of Dutra. This suggests that the GMOs were absent from the political concerns of organizations and parties throughout the 1998 electoral period. This shows that Olívio Dutra and his party (PT) had not expressed any concern for the GMOs until the second half of 1998. And to the extent that their agenda was shaped by a political alliance that embraced the MST and other rural organizations, it can be assumed that the GMOs did not present themselves as a very clear issue for these organizations until that moment either. At the same time, and this will have an effect on the debate on the GM crops later on, both Olívio Dutra and the political alliance that supported his election had in their agenda a central concern with agrarian reform based on agroecological family farming.[11]

The absence of any concern with GMOs in agriculture until this period can also be seen by the fact that the first public demonstration against these products in the state was promoted by Greenpeace, an organization that was not directly active in the regional political scene. The organization took advantage of the region's main trade fair (Expointer), in late August 1998, to publicize its campaign against the use of GMOs in agriculture. The organization thus played a relatively important role in the emergence of the conflict that began in the south of the country. The organization will participate in the event to try to convince the agricultural secretaries present at the meeting to prevent the liberation of RR soybean planting in the region. The local media will affirm that Greenpeace members would have alerted the then minister of agriculture, Francisco Turra, to the fact that Brazil was, until that moment, "the only large producer without legal transgenic crops" (Heberle, 2005, p. 212). With this, Greenpeace sought to influence the region's public authorities on the issue of GMOs by using, at the time, a legal rhetoric rather than a moral or environmental one.

Dutra took office in the second half of 1998 and did not initiate any direct confrontational actions against GM crops until March 1999. However, his first actions at the end of 1998 already indicated the government's willingness to gain greater control over GM crops. Its first measures to create a traditional (non-GMOs) grain production pole in the state, for example, were accompanied by measures that sought to strengthen control over GM crops for scientific purposes in the state. This shows that, unlike the conflict examined in the next chapter, in which issues associated with environmental risk and the scientific evaluation of RR soybean were of great importance, in the South it was commercial issues and the existing relations between small and large farmers that were more central.

In March 1999, the Plant Production Department of the Agriculture Secretariat of the Dutra government started a process of notification of research areas (from 59 to 80 trial areas) involving GMOs. On March 17 of that year, it interdicted 435 hectares of the Palmeirinha Farm, located in the municipality of Glorinha. Although

---

[11] Although the issue of GMOs was alien to political discourse in the run-up to the RR soybean release, these have gradually come to be perceived as a reality in the region, as can be seen below. With the news involving the release of RR soybean planting, these will be quickly perceived as a threat to the agrarian reform that Olívio Dutra had defended in his election campaign.

4.5 GM Seeds and Land Reform

the plantation had already obtained an injunction to harvest, the Dutra government issued a decree that demanded that the companies responsible for the plantations present documents proving that research with GMOs had been carried out (Heberle, 2005). A few days after this initiative, militants from the MST and the Movimento dos Pequenos Agricultores (MPA) held a protest in front of the Instituto Rio Grandense do Arroz (IRGA). The protest sought to pressure the state to suspend the agreement that the government had established with the company Hoechst. The protesters also intended to destroy a plantation with 300 genetically modified plants. However, the action was cancelled when the activists were "convinced that this was research" (Heberle, 2005, p. 232). This indicates that resistance to GM crops was up to this point directed exclusively at the marketing of these products, setting a precedent for acceptance, at least in this phase of the conflict, for plantations whose purpose was scientific and not economic. This will also be expressed in the bill that the government will seek to approve in order to impede the commerce of these products in the state. This will establish economic and not scientific restrictions.[12]

During the period, the Dutra government did not present a clear justification for this change in attitude towards GM crops, and although it had not taken any more direct decision to curb the planting of GM seeds until that date, there was evidence of its government's concern in achieving greater control and inspection of GM crops. This will be shown by the government's stance of denouncing the cases of imported seeds and crops that the state government will classify as illegal. In this period, the "state strategy was to denounce the existence of farms cultivated with transgenics and request a federal position" (Heberle, 2005, p. 232). This initial resistance of the Dutra government to GMOs in agriculture would immediately place it in confrontation with the federal government. In this phase, the Dutra government will show itself dissatisfied with the federal government "which was responsible for oversight" (Heberle, 2005, p. 232). What will show is that, until that moment, the state government will recognize the responsibility of the federal government for the supervision of GM crops. However, the Dutra government's position changed as of 1999, with the "Um Território Livre de Transgênicos" campaign, when the state government sought to directly assume responsibility for this process, thus contesting the federal government's authority to carry it out.

Action against GM crops becomes more forceful in January 1999. In that period, the State Attorney's Office opened a civil investigation to investigate the entry of GMOs into Brazil. The decision is the result of a representation made by deputies of the governor's party (PT) that alleged at the time the lack of control in the entry of these products in the region. The Dutra government suspected that CTNBio itself was at that time granting authorizations for importing RR soybeans. In the words of Heberle (2005, p. 228): "The Public Ministry was initiating investigations by requesting information about the import authorization for RR soybean given by the *Comissão Técnica Nacional de Biossegurança* (CTNBio) to the Monsanto

---

[12] As can be seen, these actions ended up creating a somewhat confusing scenario in the debates on the GMOs in the region. Scientific research on GMOs has, from the beginning, been linked to the interest in the product.

company," Obviously, from CTNBio's perspective, such a request was inappropriate. Since CTNBio is the institution responsible for assessing the safety of GMOs and, based on this responsibility, validating any release of these products, it could be assumed that any favorable opinion on GMOs in agriculture would be based on a reliable assessment by CTNBio itself.

The existing border between RS and Argentina, a country that was a producer center of GM seeds, made soybean imports a focus of concern in the RR soybean release period. RS has a border region of over 1500 km with countries such as Argentina and Paraguay, where GMOs were already legalized. This border region will present itself as a vast area for soybean planting that offered in the period the possibility of illegal seed trafficking in the region (Pelaez & Albergoni, 2004). One of the main bills that the Dutra government will seek to approve against GMOs in agriculture in 1999 will be justified precisely by the risks of importing RR soybeans across the border with Argentina. José H. Hoffmann, then secretary of the Dutra government, will state in 1999, when the government had already started its political campaign against the RR soybean, that: "It is not enough to curb the planting because we know that Argentina plants transgenic soybean and, among other things, exports it to Brazil in a draw-back system, that is, Argentine soybean is industrialized in Brazil and its by-products such as bran, oil and lecithin are exported" (Hoffmann, 1999, p. 173).

At the beginning of 1999, while the Dutra government sought to ban the GM crops in the state, it signaled a first approach to environmental discourse to justify its resistance to these products. Organizations like Greenpeace and IDEC opposed the federal government's decision to release RR soybean and sought, through a public civil action, to bar the release of RR soybean in the country. These organizations sought to point out the limits of the RA used by CTNBio to release the soybean and defended the use of Environmental Impact Assessment (EIA) to regulate its release. Olívio Dutra's government will proceed in a similar way by linking the possible release of GMOs in agriculture to these conditions. His government will take different measures to try to prevent the GM crops in the state, using the same arguments and strategies that Greenpeace and IDEC would use at the national level. The 1999 bill, proposed by a member of parliament from the ruling base, sought precisely to ban the cultivation and marketing of GMOs based on these arguments. In March 1999, the governor signed a decree that determined that areas carrying out research with GMOs had to notify the public authorities. It also demanded, simultaneously, the realization of EIA-EIAR for GM crops that had a strictly scientific objective. With this, the Dutra government used the same arguments used by Greenpeace to try to paralyze the trade in GM seeds in RS in the initial phase of the conflict.

4.6 A Late Threat from the Green Revolution

There are several factors that link GMOs with distributional issues in agriculture. Work examining the social and economic impacts of GMOs indicates different ways in which agricultural biotechnology can alter the distribution of costs and benefits in agricultural production. If the introduction of seeds into crops results in increased costs for planting, or new contractual burdens for small farmers, the latter may find it difficult to obtain financial resources to finance their plantations (Gonzales, 2007). It has been argued that patents linked to GM seeds can substantially alter traditional trade relations, inducing small farmers to buy, for example, new seeds each season, preventing them from stocking up on seeds for future planting. At the same time, if GM seeds were to provide the productivity increases that their advocates often envision, this could lead to a loss of competitiveness for small farmers. The latter could also suffer economic losses if the introduction of GM seeds entails a reduction in manual farm labor. Moreover, by contaminating non-GM crops, GM seeds could also inflict losses on small farmers if buyers of non-GM products restrict their purchases due to the absence of any controls to segregate GM crops from traditional crops. Small farmers could also suffer losses for more environmental reasons. Some of the characteristics associated with GM seeds could lead to limitations in adapting crops to local contexts in particular regions. The economic and technical dependencies associated with GM seeds may also lead to a loss of cultural traditions in the agricultural practices of small family farmers. In this context, concerns are less about the existential value of family farming and more about the potential of agbiotech to generate significant changes in the social lives of small-scale farmers and their rights. It has not been uncommon to see the family farm structure itself as expressing a system of rights in itself. In such cases, as Thompson (1997, p. 107) points out, it is believed that an "economic structure including both wage labor jobs and the opportunity to enter or leave family farming at will is *ipso facto* an ethically just structure of economic opportunity rights."

The concern with the impacts of agricultural biotechnology on small farmers is also linked to an older debate involving the structural change brought about by technological innovation in the agricultural space, a change that is captured by the idea of technological treadmill.[13] In English, this process is designated by the expression of a fewer and larger mechanism. The expression refers to the reduction in the number of farms (fewer) and the increase in their size (larger). The thesis suggests that the flow of technological innovation that occurs in agriculture, especially in capitalist societies, leads to a structural change in agriculture that tends to favor large property, subsequently leading to a decrease in the autonomy of small farmers. Technological change is seen as a threat to virtues when this change (a) promotes an increase and expansion of a social rationality characterized by economically formalized relations, (b) leads to a realization (performance) of these practices in a

---

[13] The technological treadmill thesis in agriculture was first enunciated by Willard Cochrane and then projected itself into Marxist-inspired works. On these questions, see Thompson (1997).

routine and unreflective manner, and (c) threatens the stock of knowledge and practices of family farming (Thompson, 1997).

When examining the conflict in the south of the country, it is clear that RR soybean was perceived as an innovation brought about by the same modernizing processes that shaped the green revolution in the past. This is due to the characteristics of the agroecological agricultural projects that were one of Dutra's agrarian flagships. Agroecology, due to its principles, tends to be in direct opposition to the GMOs and modern agriculture. The first actions of the Dutra government signaled its attempt to strengthen this type of agriculture in the state.

As seen before, these concerns were absent among advocates of RR soybean release. In the period of RR soybean release, there was no concern on the part of the MS&T, let alone CTNBio itself, to examine the economic consequences that RR soybean release might have for the most vulnerable sectors of Brazilian agriculture. Thus, the RA that guided the release of RR soybean did not include any consideration of its economic consequences on the Brazilian agricultural structure and much less on the sectors that could lose most from its release. On the other hand, some of these issues were rightly put forward by the Dutra government to bar GMOs in the state of RG. Resistance to GMOs in agriculture will include issues associated with national sovereignty, respect for farmers' culture, health, the environment, and also issues associated with the regional market. However, it is possible to verify, by the actions of the Dutra government itself, that some of these reasons were more important than others. Especially those that touch on the defense of family farming and agroecology. This can be seen by the contradictions and tensions existing in the discourse and actions of the Dutra government in the conflict. In June 1999, the Secretary of Agriculture would state that:

> Some pro-GMO scientists say there is nothing conclusive about compromising health. We have told them that this is exactly the argument for not releasing GM foods. If there is nothing conclusive about harmfulness, there is also nothing concrete about non-harmfulness. The argument that they use against us is exactly the one that we adopt with much more reason against them [...] The public agent responsible and committed to the health of the people cannot release a food if there are doubts about the consequences of its ingestion for the health of the people. (Hoffman, 1999, p. 170)

In this passage, the secretary does not express substantive conviction about the impacts of GM foods on human health and environmental risks, which does not mean that they have not received the government's attention. However, this perception is different from that which the government itself will express on the impact of GMOs on the agroecological family farming structure, as will be seen below. In this last dimension of the conflict, the risks posed by GMOs in agriculture are seen as much more threatening and real. In short, while in the dimension of health and the impacts linked to it, the Dutra government will acknowledge that "there is nothing conclusive about harmfulness," thus assuming doubt for this issue, the Dutra government will take the impacts of GM seeds on the interests of small farmers as more real and worrying.

In this context, the possibility should not be excluded that even concerns about the health and environmental impact of GMOs in agriculture may then be linked to land reform issues in some way. At the international level, trade barriers placed by governments on certain products can be justified by the need to manage certain risks associated with technological advance. And, once this is seen as being in the public interest, this possibility is provided for in the WTO Sanitary and Phytosanitary Agreement, which states that in the absence of scientific evidence as to the safety of certain products, "its member countries may adopt sanitary and phytosanitary barriers for the purpose of preserving the environment and public health" (Pelaez & Albergoni, 2004, p. 206). In this case, the Dutra government's measures could eventually be inspired by these legal precepts to prevent the commercialization of GM seeds, finding additional arguments, even if speculative at first, to oppose to these products in the region. However, in the Brazilian context, it is the federal government, and not municipalities or regions within it, that have the power to activate the laws established in these agreements. However, with a possible change in the political scenario, and a future federal government aligned with the fight against GMOs, the construction of this discourse could prove politically and strategically important in the unfolding of the conflict. Thus, if any such moratorium could become relevant for the resistance to GMOs, environmental and health issues could become relevant for Dutra's political campaign in the future, since they could justify the establishment of an economic barrier through arguments related to public health. In this context, issues involving health and environmental risks could set a precedent for the construction of a future trade barrier to GMOs in the state.

This is because, under international agreements, the possibility of creating any type of economic embargo on certain products needs to be linked to this type of issue, which excludes concerns involving distributional impacts on small producers. The conventions and the biosafety protocol, for example, do not set a precedent for countries and regions to create economic embargoes on GMOs based on distributional issues involving small farmers. On this point, Gonzales (2007, p. 29) notes that it would not be possible for poorer countries "to reject GMOs based on socio-economic considerations not directly related to impacts on biodiversity, such as harm to the livelihoods of local and indigenous communities or increased dependence on proprietary seeds and other inputs produced by transnational corporations."

4.7 Fiscalizing (Il)Legal GM Crops

On March 3, 1999, Dutra will declare RS an "Um Território Livre de Transgênicos." The decision was not only political, as a result of the government's campaign against GMOs in agriculture, but also enunciated itself, in a way, as a type of legal action. The aim was to make the campaign operate as a type of regional moratorium that would be equivalent to the moratorium initiated in Europe in the same period. Thus, the decision was accompanied by the signing of Decree No. 39,314, which was intended to regulate State Law No. 9453 (a law dated 10/12/1991), which required the notification to the government of actions involving research, tests, experiments, and other activities in the areas of biotechnology. In addition to requiring the notification of GM crops, the decree also required the submission of an environmental impact study and report (EIA/EIAR). In parallel to these actions, the Legislative Assembly of the State of Rio Grande do Sul was passing the Bill No. 016, created on February 17, 1999, authored by the PT deputy Elvino Bohn Gass, which sought to prohibit the cultivation and sale of GM foods in the state (Almeida & da Silveira, 2000). Therefore, the government sought to create legal means to prohibit the planting of RR soybean in the region. On the other hand, in the same period, other PLs that sought to do the opposite, to establish the conditions for the release of RR soybean in the state, began to be processed in the state's Legislative Assembly. Thus, the same legal strategy that Dutra sought to implement in the state government was reproduced in the following months in several municipalities in the RG, pursuing the opposite objective from the one expressed by the Dutra government. Several of them went on to pass municipal legislations either prohibiting or approving the marketing of GM foods in their municipalities.[14]

However, many of the legal actions that Olívio Dutra tried to implement in 1999 against the release of GMOs did not find political support in the state legislature. The laws that the governor sought to institute to implement his plan for a "Um Território Livre de Transgênicos" were simply not passed by the legislature. And, even if these laws did pass, it must be considered that they would clash with federal legislation, setting a precedent for a legal clash between the state of RS and the federal government, since legislation and decisions on GMOs are confined to the latter. In fact, one of the reasons that these bills gained little support in the state legislative house is linked to the recognition by members of the assembly that they lacked legality. At least, this was the justification offered by some of them for rejecting the bill that would be proposed in 1999 by the Dutra government. It is important to return here to the legal initiatives that were proposed by the government and that were rejected by the Chamber of Aldermen. Despite their failure, they enunciate the intentions of the government period. The bill proposed by the government had in its articles the following guidelines:

> Article 1 – The commercial cultivation of genetically modified organisms (GMOs) in the State of Rio Grande do Sul is hereby prohibited.
>
> Sole Paragraph – For the purposes of this law, the definition of GMO expressed in Articles 3 and 4 of Federal Law 8.974 of January 5, 1995 shall be considered.
>
> Article 2 – The commercialization of products containing in their composition substances originated from genetically modified organisms intended for human or animal feeding is hereby prohibited.
>
> The companies that develop or intend to develop researches with genetically modified organisms in the State of Rio Grande do Sul shall report their activities pursuant to Law 9453, of December 10, 1991.
>
> Article 4 – This law shall enter into force on the date of its publication, revoking provisions to the contrary. (Rio Grande do Sul, 2014a, s/p)

---

[14] Examples of municipalities that tried to ban it are Espumoso (Dec/98) and Pelotas (Oct/99) and examples of municipalities that tried to support the release are Cruz Alta (July/99), Jóia (Sept. and Nov./99), Tupancieretã (Sept. and Nov./99), Não-Me-Toque (Sept. and Nov./99), and Redentora (Sept. and Nov./99) (Rede Brazilian Sustainable Agriculture, 2000).

4.7 Fiscalizing (Il)Legal GM Crops

The project, as the Secretary of Agriculture will indicate, "was elaborated with the agreement of the executive power" (Hoffman, 1999, p. 174). The secretary will also justify the creation of the project in the period with the following arguments:

> We also have a bill in process in the Legislative Assembly that foresees the prohibition of production and commercialization of transgenics in our State. The measure is not restricted to soybean, but it is extensive to all transgenic activities. The project is very important, because it is fundamental that the commercialization is also prohibited. What we have done up to now is to apply Law no. 9.453, which makes a series of requirements for carrying out these experiments that, for the time being, have not been met. (Hoffman, 1999, p. 174)

The bill was based, in turn, on a law enacted by the state in 1991 which, in its article 3, stated the following:

> Article 3 – Companies that develop or will develop research with genetically modified organisms in Rio Grande do Sul shall report their activities pursuant to Law 9.453, of December 10, 1991. (Rio Grande Do Sul, 2014b)

The state did not have any specific structure to organize this process. As noted above, GMOs were not a tangible agricultural reality until the end of 1998. On the other hand, this bill worked with an unconditional prohibition of these products in the economic sphere and, at the same time, opened a precedent for their liberation in the scientific research sphere. In other words, it established a commercial prohibition while creating conditions for planting for scientific purposes. Law 9.453, which is mentioned in article 3 of the bill, stated, in turn, the following:

> Article 1 – Domestic or foreign companies that carry out research, tests, experiments and other activities in the State of Rio Grande do Sul in the areas of biotechnology and genetic engineering involving Genetically Modified Organisms (GMOs), as well as products resulting from such technology, shall notify the Executive Branch as provided for in this Decree. (Rio Grande Do Sul, 2014c)

The Secretary of Agriculture of the period himself recognizes that, until that moment, there was no specific law in the state that could prohibit the trade of GM seeds. And to the extent that the Dutra government had failed to approve the bill banning commercialization, it would have as its only trump card Law 9453, which as such makes notification requirements for activities involving scientific research with GMOs. However, this law did not establish any prohibition on trade. This shows that, as of the RR soybean release, the state had no legal safeguards to prevent the trade of RR soybean in the state, other than those existing in the biosafety law itself and in Law No. 9453. Which indicated an uncertain and fragile picture for the government to legally sustain the ban on GM food trade in the period. After all, CTNBio itself would justify the release of GMOs in agriculture based on the biosafety law, which, as such, was legally above any law that the Dutra government could enact.

This legal framework will have important consequences for how the conflict unfolds. With the release of RR soybean by CTNBio, many farmers have bought GM seed not for research, testing, or scientific experimentation, but for the sole purpose of conducting an economic transaction for a product that has been legalized by the government. This issue is especially important because if it is considered that many farmers started using RR soybeans for strictly economic purposes, it would mean that the Dutra government could not use Law 9453 to frame those farmers, since this law was only meant for cases involving scientific research. Thus, if there was no longer a research purpose for a product that was now legalized, no inspection of RR soybean could be carried out. If before the release period these crops could be classified as "scientific," a change occurs from the moment their use was cleared by CTNBio. They also, and perhaps fundamentally, have a strictly economic purpose.

The draft law that Dutra sought to pass in the state had further contradictions. The bill intended to provide conditions for strictly scientific GM crops, while at the same time establishing restrictions on their commercial use. However, this would not be possible without the project itself presenting problems, since the scientific research carried out by the companies is done with the possibility of commercializing the product in mind. Therefore, what would be the utility of a strictly scientific release of the product without simultaneously opening a precedent for later economic release? In the ambit of the biosecurity law, for example, the conditions for scientific research are integrated with the possibility of later commercialization of GM seeds. To the extent that research attests to the productivity and safety of seeds, this same research opens a margin for the future commercialization of GM seeds. This position of the Dutra government points to an important aspect of the conflict. The objective of his government was not to establish measures to promote the environmentally safe commercial release of GMOs in agriculture, but to prevent any commercialization of these products in the state. But in attempting to establish a ban in strict terms on GM seeds, the bill came in clear opposition to the federal biosafety law itself. In any case, the state's room for maneuver to establish legal impediments to GMOs in agriculture was practically nonexistent, unless it sought to do so by establishing a conflict with the biosafety law.

The attempt to promote an unrestricted commercial ban on GMOs in the state seems to suggest a reason for resistance to these products that transcends concerns about environmental risks. This resistance seems to have its origin in the formulation of the legal guidelines that have come to guide agricultural policy in the country and that takes us back to the very Cultivar and Patent Law that preceded the biosafety law itself. As the agriculture secretary of the Dutra government will say:

> When the National Congress quickly approved the Law of Patents and the Law of Cultivars, it renounced Brazilian sovereignty in seed production and, consequently, food production itself. During his first term, President Fernando Henrique Cardoso sent to the National Congress and it approved these laws. From then on, the whole process began. In reality, we are discussing a first consequence of these laws, that is, we are being victims of the legislation, at the same time that we are awakening to what they represent. (Hoffmann, 1999, p. 168)

The Secretary of Agriculture locates the origin of the conflict involving GMOs in agriculture in the promulgation of the Cultivar Law created during the Cardoso administration. This law would become one of the main regulatory frameworks for defining the ownership of agricultural products in the country. Indirectly, it places the issue of seed ownership as the central dimension through which the conflict will begin. The Plant Variety Law, while enunciating some laws that would apparently

4.7 Fiscalizing (Il)Legal GM Crops

be convergent for the agricultural structure of small farmers, incorporates other guidelines that could be seen as posing various risks and losses for them. Article 10 of that law, for example, states that the property right over the cultivar would not be violated in circumstances in which the farmer:

I- reserve and plant seeds for their own use, in their establishment or in the establishment of third parties whose possession they have; [...]
II- uses or sells as food or raw material the product obtained from its plantation, except for reproductive purposes; [...]
III- uses the cultivar as a source of variation in genetic improvement or scientific research; [...].
IV- being a small rural producer, multiplies seeds, for donation or exchange, exclusively for other small rural producers, within the scope of financing or support programs for small rural producers, conducted by public agencies or non-governmental organizations, authorized by the Public Power [...]. (Brasil, 2005, s/p)

However, the most controversial point of this law, and the one that seems to be directly linked to the conflict that occurred in the south of the country, is in the paragraph that establishes the need for authorizations for those who develop cultivars from others already protected by the law. In this part of the law, authorizations are required in the circumstances that:

I- repeated use of the protected cultivar is indispensable for commercial production of another cultivar, or hybrid, the titleholder of the second cultivar is obliged to obtain an authorization from the titleholder of the first cultivar's protection right;
II- a cultivar is characterized as essentially derived from a protected cultivar, its commercial exploration will be conditioned to an authorization from the titleholder of this protected cultivar. (Brasil, 2005, s/p)

It is these aspects of the Cultivar Law to which the secretary is most likely referring. On the basis of these guidelines, if small farmers were to use a legally protected cultivar, or were to show the ability to extract cultivars derived from that matrix, they would have to "obtain the authorization of the holder of the protection right" (Brasil, 2005, s/p). It is for this reason that Hoffmann will see the release of RR soybean as bringing important distributional issues. These distributive issues would already be present in the Plant Variety Law, as he suggests. If this law presented an economic and cultural risk for small farmers' agricultural practices in the legal sphere, RR soybean will represent a risk in practical terms. The release of RR soybean would be an application of the Cultivar Law that he is critical of and considers unfair. Hence, the government's action did not move in the direction of improving and strengthening regulatory instruments so as to provide a secure basis for the commercial release of GM seeds, but simply sought to prohibit their commercialization in absolute terms. In part, this was because any measure that might make RR soybean environmentally safer would not eliminate the economic inequalities that were being perceived by the Dutra government and other groups opposed to the ban. It is as if economic inequality was already inscribed in the Cultivar Law itself that regulated the commercial release of GMOs in agriculture. And it would be on this dimension that the government's concerns were more centrally directed.

At the end of 1999, the Legislative Assembly approved Bill 214/99. This law was authored by Deputy Frederico Antunes and sought to repeal State Law No. 9453 and Decree No. 39,314 that regulated it. The bill recognized the role of federal agencies to approve and supervise GM crops, requiring only that farmers or companies communicate to different state government agencies (Secretariat of Health, Secretariat of Agriculture and Supply, and Secretariat of Environment) what they were doing. The bill is seen as a way of giving the state government strict responsibility for meeting the "federal requirements for the release of genetically modified organisms" (Deputy Frederico Antunes, 1999 apud Almeida & da Silveira, 2000, p. 7). Bill 214/99 sought to make clear the state government's responsibility in the process of releasing GMOs, leaving these same responsibilities in line with those existing at the federal level. Olívio Dutra had, for this reason, to veto the project, his own veto being overturned afterwards in the Legislative Assembly. Some interpretations of the episode suggest that Dutra would not have implemented his bill because his political base was a minority in the state Legislative Assembly. However, the approval of Deputy Frederico Antunes' bill was justified not by the defense of the GMOs, but by the need to make the state's laws simply conform to those of the federal government.

Although our considerations indicate the reasons for the Dutra government's resistance to GMOs in agriculture and help us understand part of the factors that shaped the conflict in southern Brazil, these elements do not allow us to understand some quite important facts that occurred in this conflict. Much of what was examined in this part of the work suggests that the conflict over GMOs in agriculture was obviously not only a conflict over the environmental risks of biotechnology. Part of the elements that shaped this conflict must be sought in the context surrounding the debate over agrarian reform in RS. Pelaez and Schmidt (2000), in examining the conflict, argue that it can be seen as representing "a continuation of a process, initiated at the end of the 1970s, of questioning the agricultural model adopted in the country and combating regulations that are favorable to international industries supplying inputs, machinery, and equipment". This is what can be understood from the Dutra government's discourse in seeing the release of RR soybean as a kind of late "green revolution."

Thus, it is not uncommon for the conflict in RS to be viewed with a certain perplexity, even by researchers familiar with the local reality. This is due to the specificities associated with it when compared to other similar cases, such as the French case:

> In both cases, the governments propose to enforce the law; however, the two cases differ in terms of the reactions to governmental actions that seek to enforce the law. While in Rio Grande do Sul the inspection action of the state government provokes public manifestations and opposing actions from pro-transgenic soybean producers, when the French government orders the destruction of "accidentally" contaminated rapeseed crops, it does not seem to find any resistance in society. (Menasche, 2002, p. 217)

This consideration assumes an existing legal pattern to the conflict. Both in France and in the case of RS, the governments were trying to "apply the law." However, the result of the judicial processes linked to the actions of the Rio Grande do Sul government demonstrated that the legal actions of the Dutra government had been invalidated. The same will occur with its agrarian reform policy, which, in the same way, also faced insoluble legal limits. In the context of the commercial release of

RR soybean, to take for granted the thesis that Dutra was trying to "apply the law" simply distorts some of the essential elements of the conflict. It overlooks the fact that the actions and decisions of both the government and the pro-GMOs farmer were, throughout the conflict, under the sieve of an ongoing legal conflict.

Both agrarian policy and biosecurity policy, areas in which the Dutra government will seek to plead for decision-making autonomy, are matters of responsibility of the federal government. Therefore, in this passage, the two "governments" are not equivalent things. In Brazil, the body with the greatest responsibility for decision-making on the release of GMOs was, and still is, the federal government, which created the conditions for the release of RR soybean. For their part, in France and in Europe as a whole, the decision was in favor of declaring a moratorium on GMOS in 1999. This did not exist on the part of the federal government in Brazil. Therefore, when examining the government's political stance, taking into perspective what has happened in Brazil and Europe, very different stances are perceived due to the fact that the governmental levels are also distinct. Had Dutra been the president of Brazil in that period, it might have been possible to find a similarity between the cases. It was precisely because of this scenario that the Dutra government needed to engage in a legal conflict with the federal government in order to try to obtain decision-making autonomy over the release of RR soybean. Even if existing GM crops in the south of the country were considered as the fruit of an illegal action, that in itself would not give the Dutra government any of the authority it sought to confer on itself. This legal confusion occurs precisely because of the judicialization of the conflict, which tends to generate instant legalities that are soon undone by court decisions that follow one another depending on the speed of the legal dispute in the country. Thus, would the decisions of the Dutra government be legitimate under the law when considering the decision of Judge Souza Prudente, the judge who allowed the GM crops to be considered illegal? Or should they be evaluated based on the decision of Judge Selene Maria de Almeida, who offered an opposite interpretation soon afterwards?[15] Or should they be evaluated based on the final decisions of the STF, which provided a legal basis for CTNBio's decisions?

Furthermore, even if an illegality is attributed to the GM crops in the period, it is not certain that it would be up to the Olívio Dutra government to inspect them, since, as already noted, the biosecurity law gives the federal government this responsibility, an allegation that, by the way, was raised by pro-GMO farmers throughout the conflict. Hence, the RS case differs from the French case. In the latter, the decisions emanated from the higher instances of power that, as in Brazil, are responsible for regulating GMOs. In the RS case, the decisions emanated from the (RS) local government, which has never been recognized, either by the pro-GMO farmers, or even by the courts, as the legitimate sphere to regulate these products in the region. The

---

[15] Judge Souza Prudente's decision allowed the classification of GM crops existing in the country as illegal, since, according to its vision, the soybean release did not follow the legal requirements which are provided for by the environmental biosafety law. However, the decision of that judge was annulled by Judge Selene Maria de Almeida in 2002, which, in opposition, would establish as legal the same plantings.

Dutra government lacked, therefore, any legal and political authority to decree a moratorium on GMOs as had occurred in Europe and France in particular. Thus, in a retrospective analysis of the conflict, it would not be inappropriate to consider the position of the pro-GMO actors as a struggle to defend a legality that was later imposed on the country by the Lula's own government. A government that Dutra himself, ironically, would support in the electoral campaign and that ended up implementing a political agenda for the GMOs that was the opposite of that promised by the governor himself. This explains the end of the confrontation of the local government (RS) with the federal government already at the beginning of Lula's government.

Therefore, one cannot overshadow the fact that the very legality of GM crops was an important element of the conflict. Once released by CTNBio, RR soybean came to be seen as a legalized product to be marketed by farmers. At the same time, the importation of soybean from Argentina allowed farmers to deviate from the laws established by the state government, which, as already mentioned, turned to scientific research activities. Once RR soybean was released by CTNBio, the soybeans used by farmers were imported and their use in plantations had no scientific purpose. Its purpose became exclusively economic in the eyes of the farmers. Therefore, since the farmers were no conducting any "scientific tests" with the RR soybeans (what they never did), why should they notify the state public power? That is why the veto to commercialization was crucial for the Dutra government, since state legislation did not offer any impediment for this type of situation.[16] Thus, considering this whole picture, could the Dutra government have classified the GM crops as cases involving scientific tests that had not been notified to the public authorities? As the Dutra government intended to do? This question can be considered valid even after Judge Souza Prudente's decision made the planting and trading of RR seeds illegal. This is because the illegality of many plantings did not stem from the development of any kind of scientific research, but from the importation of an agricultural product that had previously been legalized by an agency (CTNBio) of the federal government itself.

In fact, few, if any, farmers in RS had any intention of planting the GM seeds with the objective of conducting "scientific test" after the soybean was released by CTNBio. The farmers' objective was effectively commercial in nature. The "test" for the farmers was more related to issues involving resistance and productivity than environmental safety and was not associated with the possible environmental risks that they were unlikely to be able to examine. Or as Bauer (2006, p. 237) will put it: "Many farmers in Rio Grande do Sul planted some areas with transgenic varieties and others with conventional seeds, testing gains for productivity and drought resistance". The test that many farmers were doing, therefore, was an economic test of

---

[16] Let us remember that the existing laws that supported Olívio Dutra's actions were directed for research rather than commerce. The government's bill was precisely in the interest to curtail the commercialization of RR soybean (and GMOs in general) in the state, precisely because the laws in force were limited to discriminating actions involving scientific research. But this only could be produced by clashing with federal law.

productivity. And not a test associated with the possible environmental risks of RR soybean.

The possibility of classifying the plantations as "scientific research" was essential to the Dutra government, as only in this way could the government grant itself minimal regulatory power over these plantations. However, the government encountered several obstacles in pursuing this goal. In many cases, the seeds had been obtained, as already indicated, through a commercial import process. With the release of RR soybeans by CTNBio, therefore, the possibility of importing them was created. This is because, once approved, RR soybean was treated as equivalent to other soybeans. Both for the federal government and for the farmers who imported it. There was, for these cases, no illegality in importing GM seeds. Which means that, in the latter case, the Dutra government would not have the right, taking the perspective of the pro-GMO farmers, to carry out inspection on plantations of a soybean that had previously been legalized by the federal government. A product that ended up being acquired through a commercial negotiation (the importation of the seeds) and not through "scientific plantations" that should have been notified to the government. The legitimacy of the state to inspect them was unclear, then, to the pro-GMO farmers, since their decisions were economic and occurred within a framework of legality promoted by the federal government.

The legal uncertainties that permeated the Dutra government's actions to supervise the GM crops are important for us to consider some aspects of the conflict. Although the Dutra government established a policy of opposition and confrontation with the GMOs in agriculture, this position is not sufficient to understand some of the important tensions that arose in the period between the state government and the large farmers in RS. For this dimension to be better understood, a return to the agrarian conflicts that involve the debate on agrarian reform in the region needs to be carried out. How it will be possible to realize, these conflicts projected themselves into the case involving the commercial release of RR soybean, making the issue of GMOs and the issue of agrarian reform inseparable in the local conflict. Although the facts reported below may seem totally strange at first, they will make more sense when the RR soybean conflict is examined in the light of this issue.

The analysis of the bill examined above offers us clarification on the aspects involved in the justification that the Dutra government would present in its opposition to the GMOs. By examining some of these points, it is possible to observe that the bill defended by Dutra was not dissociated from his agrarian reform policy. His political platform for agriculture was characterized by a commitment to agroecology and family farming. In the election campaign period, Olívio Dutra (1999 apud Heberle, 2005, p. 229) would then declare that: "The Popular Front government will encourage ecological agriculture, stimulating the organization of ecological farmers' groups by guaranteeing credit, technical assistance and improvement of these farmers with training courses."

The Popular Front had a political profile of open opposition to the federal government's agrarian policies, transforming the demands of agrarian movements linked to family farming and agroecology into the government's political agenda. In turn, these guidelines were directly associated with distributive issues in agriculture, since they addressed issues involving the autonomy of small farmers. Although the issue of GMOs was alien to the political debate that brought Dutra to the government, these would quickly be recognized as a threat to the agrarian reform that had become Olívio Dutra's (PT) main campaign banner. This political composition of the government would have a relevant symbolic and political effect on the RR soybean conflict, since it indicated a strong commitment of the government to these movements and their proposals for agricultural policy. What permeated this close connection between the Dutra government, the MST, and other organizations linked to small farmers was a shared vision for the creation of a differentiated agricultural policy in the state. The campaign for an "Um Território Livre de Transgênicos" was, in a certain way, part of the agricultural proposal that Olívio Dutra had presented during the election period. As Almeida and da Silveira (2000, p. 5) recall: "the determination of the 'free zone' represented one of the first steps in the consolidation of a proposal—presented by the PT during the electoral campaign—that prioritizes family farming."

It should be remembered that it was precisely in RS that organizations associated with family farming were developed in Brazil as movements to defend the autonomy of small farmers. Already in the second half of the 1970s, several organizations and groups in southern Brazil sought to rescue the idea of the "family farmer" with the objective of "reorienting the productive systems and the technologies employed towards strengthening the economic capacity and autonomy of this category" (Almeida, 1999, p. 58). The idea of autonomy will become one of the main ideas of the set of conceptions and practices of alternative agriculture (agroecology) in environmental and agricultural organizations in the region.[17] Thus, for the MST, the valorization of family farming is generally associated with an image of "the return of idealized living conditions to a good past, of autonomy, prior to the subordination of the farmer to multinational industries" (Veras, 2005, p. 39). Criticism and resistance to GM seeds are inseparable from the imaginary that these organizations' structure was based on ideas and practices involving family farming and agroecology. The threats that GM seeds seem to pose to family farming may involve a sense of loss that brings us back to an existential value linked to this type of social structure in the agricultural system.

When examining the conflict, it is necessary to remember that, although the MST recognized family farming and agroecology as flagships of its actions, the defense of this type of agriculture occurred at the beginning of the 1990s only in the discourse of movement leaders and in campaign primers. This vision of these alternative agricultural models was nonexistent in the movement's discourse and practice before that period. Throughout its existence, rather than seeing modern agriculture as a threat, the MST sought to imitate and even promote it in the settlements it created. In these settlements, the goal was to implement large-scale modern agriculture and integrate farmers into the national and international markets. However, this

---

[17] The disputes involving the labeling process in the country also had as a background issue autonomy. However, for this case, the central issue was the autonomy of the consumer and his rights. On this point, see Chap. 6.

4.7 Fiscalizing (Il)Legal GM Crops

defense of modern agriculture was not based on a technological basis, but on an economic one. The criticism of the large latifundia occurred because of the inequality of power and wealth associated with it and not its technological model. In fact, as Veras (2005) points out, the technological question was seen as marginal for the MST until the 1990s. This meant that, rather than simply rejecting agroecology and small-scale farming, the movement perceived these ideas as "reactionary." For this reason, the news that eventually emerged throughout the conflict about GMOs, indicating that the MST itself would be planting RR soybeans in the state, should not surprise when these factors that influenced the history of the movement are considered.

The MST's position changed at the end of the 1990s, due to the ongoing failure to introduce economy of scale practices in the settlements. This led a small group within the MST itself to return to the initial ideas linked to the so-called alternative technologies. This change was influenced by the action of NGOs and consumer groups that, in the words of Veras (2005, p. 51), brought "new arguments about the 'best' technological format for the settlements." This movement gave rise to the first group of MST settlers in favor of implementing alternative technologies in MST experiments. But this return was only partial, as the movement maintained its goal of encouraging economies of scale in agriculture.[18]

In any case, this deviant behavior by MST groups did not prevent the promotion, from the 1980s onwards, of a series of events in RS that brought together different organizations linked to the countryside and focused on criticizing the growing dependence on agricultural inputs produced by large corporations. In these events, there was already a concern about the purchase and sale of agricultural seeds and the recognition of their strategic role in strengthening the creation of alternative agricultural models. These themes came to be incorporated into the MST political agenda at the beginning of the 1990s with the conference "Land, Ecology and Human Rights." In this period, the movement was strongly influenced by the theses presented by FAO/INCRA, which pointed to family farming as an impetus for a more equitable economic model. This perception was already present among organizations linked to small farmers in the 1980s and this new agricultural vision will return in the conflict over the release of RR soybean in MST messages:

> The other impact in the economic field is that the planting of transgenics generates a monopoly of seeds by multinational biotechnology companies. That is, the production, cultivation and commercialization of the seed come under the control of multinational and transnational companies whose only interest is to increase their profits. (MST, 2006, s/p)

Another passage with a similar message is the following from J. Lutzenberg's view:

> What is happening today in what was initially called—still somewhat honestly—genetic engineering and which they now prefer to call "life sciences" is a terrible conspiracy that is

---

[18] Although part of the movement can be said to have adhered to a technological paradigm based on family farming and agroecology, it is quite possible that certain sectors of the movement to remain faithful to the modern agricultural model based on agrotoxics and, recently, on agricultural biotechnology. On RR soybean planting by the MST, see Gerchmann (2003) and Rocha (2012).

the culmination of the process of expropriation of the farmer, a process that has been worsening for decades [...] The target that these and other transnationals pursue is to ban all use of the farmer's own seed. (Lutzenberg, 2000 apud Almeida & da Silveira, 2000, p. 13)

Beginning in the second half of the 1990s, MST discourses and reports placed increasing emphasis on the role of family farming in achieving "an economic model that is both more equitable (in income distribution) and more efficient (in cheaper food supply)" (Wilkinson, 2004, s/p). This vision would be incorporated by the MST in the 1990s, which began to give greater credibility to family farming itself and also to agroecology. This interest would later be incorporated into Olívio Dutra's government program. Part of the defense made of family farming in the RR soybean conflict arises from the virtues that are usually associated with it. Therefore, it has become common to see family farming, especially when integrated with agroecology, expressing ecological virtues absent in the agroexporting model represented by agricultural biotechnology. In the view of a social scientist like Henrique Leff (2002, p. 39), the agroecology incorporates the ecological functioning necessary for sustainable agriculture and, at the same time, incorporates equity principles into the production process, in such a way that its practices allow for equal access to livelihoods.

The liberation of RR soybean will occur, therefore, precisely at the moment when the movements for alternative agriculture were gaining strength with the election of Olívio Dutra in the period. The position against the planting of GM seeds came to fit in with the Dutra government's attempt to signal its independence from multinational companies, while at the same time seeking to respond "to the expectation of valuing family farming and agroecology, which became the Secretary of Agriculture's flags of struggle" (Heberle, 2005, p. 272). That is why the state Secretary of Agriculture, José H. Hoffmann, will state in the process of resistance to the release of RR soybeans that:

> Our government proposal, widely discussed during the election campaign, prioritizes family farming. Thus, we cannot agree with a technology that precisely excludes small family farmers. This is a very strong reason for our opposition to the production and commercialization of transgenics. (Hoffmann, 1999, p. 170)

In the conflict in the south of the country, it is possible to say that RR soybean was seen as expressing a line of continuity with the technological innovations brought about by the "green revolution" which, likewise, had always been perceived as a means of strengthening the power of large corporations. GM seeds were perceived as an obstacle to the new agrarian reform agenda that had been announced by the governor during the election period. The main consequence of their release would be the increasing concentration of (intellectual) property in the countryside and therefore of income and wealth. The MST perception that "the production, cultivation and commercialization of the seed come under the control of multinational and transnational companies whose only interest is to increase their profits" (MST, 2006) is merely the expression of concern, well rooted in the movement, about the social inequality produced by the innovations brought about by the large corporations. Hence, the

4.7 Fiscalizing (Il)Legal GM Crops

rejection of RR soybean is also perceived, as indicated by the Secretary of Agriculture's words, as an extension of the "green revolution" that took place in the past:

> We are suffering the consequences of the so-called green revolution, adopted with more intensity after the military coup, which boiled down to the intensive use of agrochemicals, disrespect for nature and the health of workers in the countryside of the city. (...) If we adopt transgenics we will be taking a step in the dark, a step forward in the green revolution. We do not agree with this technological paradigm, with this agriculture based on the green revolution. On the contrary, the Democratic and Popular government is joining the critical mass to quickly spread a new technological paradigm, based on agroecology, that respects nature, the health of the rural worker. We do not accept a step forward, a much riskier step than the green revolution, whose disastrous consequences we know. (Hoffmann, 1999, p. 169)

Therefore, the most important threats brought by GMOs in agriculture were not perceived, fundamentally, only for their environmental implications, but to the obstacles that this agricultural innovation could bring to an alternative agrarian reform agenda. GMOs were seen as impediment to the construction of a more inclusive and equitable agricultural model in the state of RS. RR soybean was perceived as a threat to the rights and the very autonomy of small farmers, elements that were part of that same political vision. In the period of the release of RR soybean, Hoffmann will make statements that reflect this point of view in even more detail. With the liberation, the very cultural identity of small farmers would be in question. Hence, one of the strong reasons for resisting GMOs in agriculture lies in:

> [...] respect for the historical culture of the family farmer. We cannot agree that they have to buy seeds every year, losing autonomy and independence. Historically, family farmers, who predominate throughout the country, especially in the southern half of Brazil, have traditionally set aside part of their harvest for seeds for the following harvest. However, now with the patenting of seeds, he will be obliged to buy new seeds every year. The multinationals have been improving this process of domination of the farmer. First they force the farmer to sign a contract committing him to use only part of his production for seed. (Hoffmann, 1999, p. 169)

Although Hoffman does not use the term rights, his discourse tends to invoke them when defending the autonomy and economic independence of small farmers. RR soybean posed the risk of the loss of traditional practices associated with family farming. This view reflects several concerns around the social impact of GMOs on family farming that have been examined in the previous pages. In a sense, RR soybean would be representative of the capitalist technological treadmill mentioned earlier inducing a shift that would tend to strengthen large farms at the expense of the autonomy and economic independence of small farmers. With the new soybean, these farmers would have to assume new commercial relations, being forced to buy new seeds every season. Hoffmann's vision also presumes the emergence of new contractual burdens for these small farmers, since they could no longer make use of a traditional agricultural practice (stocking seeds) that would significantly reduce production costs. The economic and technical dependence associated with RR soybean would, in turn, involve the loss of a cultural tradition that could be seen as constitutive of family farming itself ("the historical culture of the family farmer").

In the document produced by AS-PTA, an entity that supported the Dutra government, entitled Farmers in the Context of the FAO the International Treaty on Plant Genetic Resources for Food and Agriculture, this kind of rights-based interpretation is even more clearly expressed:

> The concept of "farmers' rights," as discussed in various international forums, has not yet been literally incorporated into the debate of rural organizations in Brazil. However, it is clearly expressed and embodied in the organizations' understanding that seeds constitute both material and economic resources and cultural assets that are part of the heritage of farming peoples and a condition of their very existence. This understanding of seeds as cultural assets highlights the inextricable relationship established by farmers between their knowledge and biodiversity resources. (Fernandes, 2010, p. 2)

In the specific context of seed marketing, AS-PTA states that organized rural movements are unanimous in recognizing the right of small farmers to "produce, market and exchange their seeds" (Fernandes, 2010, p. 3). Among the rights that the GMOs would be threatening would be, for example, the basic right "to be free of pesticides and transgenics" (Fernandes, 2010, p. 3). Action that could be located, as seen above, in the perception of these actors, in the Plant Variety Law itself. AS-PTA also sees in the actions of the state and multinational corporations a threat to these rights, since these actions would be leading to a "modification of regulatory frameworks aimed at protecting the commercial interests of private groups and the introduction of technologies such as transgenic seeds" (Fernandes, 2010, p. 5). It is the changes in these frameworks that would lead to "an increasing privatization of genetic resources and their monopolistic exploitation through different mechanisms of industrial protection, such as patents in the case of transgenics" (Fernandes, 2010, p. 5). And by doing so, these actions would lead to a curtailment of the traditional practices of family farmers and, in particular, practices involving, says AS-PTA, the "management, production, use, conservation, commercialization and exchange of seeds" (Fernandes, 2010, p. 5). In short, the discourse that the Dutra government professed on behalf of the autonomy of small farmers could easily be found in those agricultural organizations that gave it political support. In the discourse of these organizations, the GMOs in agriculture represented a cultural and economic exclusion. They would constitute threats to the rights of small farmers and, therefore, factors promoting a picture of environmental injustice in the region. Hence, the release of RR soybean involves distributional issues that were expressed by this axis of the conflict.

It should not be forgotten that GMOs have also represented, in this whole context, a threat to the creation of a distinct agricultural market that Dutra was targeting:

> We have been saying everywhere that the country is throwing away the unique chance of being Europe's supplier, obtaining differentiated prices for non-transgenic foods. This is not only true for soybean, but for all the others. Evidently, what really interests us is to free the consumer from transgenic foods. [Many say that the position assumed is nothing more than an ideological question. However, the existence of a differentiated market is a technical and not an ideological question. (Hoffmann, 1999, p. 172)

In addition to being a matter of principle, opposition to RR soybean also indicated a strategic market position. The creation of a market for (non-transgenic crops) conventional products could be seen here as important for strengthening the government's promised agricultural reform. Broadly speaking, favoring trade in GM seeds would be tantamount to nullifying the economic and legal conditions that would provide the competitive advantages for non-Tg agricultural products that would underpin the production of the region's agroecologically based family farming that the government sought to foster. It should be included here perhaps the fear that the impossibility of establishing any control over the segregation of plantations (Tg and non-Tg) would result in a loss of demand for the domestic market for traditional soybean, a problem that required a choice to be imposed, as the Dutra government attempted to do, on the state's entire agricultural system, preventing farmers who wished to incorporate GM seeds into their crops from doing so.[19]

Environmental groups linked to small farmers will help foment the campaign against GMOs, giving greater legitimacy to the government's actions. The European Moratorium will do the same by suggesting that, in economic terms, Olívio Dutra's actions sought to apply in his state. At least, European support began to provide a market justification for the Gaucho government's resistance. His vision was not against the market, but associated with a vision of the economic segmentation of this same market on a world scale. This means that his policy could eventually present certain points that could please the large farmers, if they were interested in the European market that Dutra was trying to approach. Should this become feasible, it would be possible, perhaps, to make conventional soybean attractive to large landowners. At the same time, it was for this reason that, when he took office, the Secretary of Agriculture, José H. Hoffmann, indicated that the commercialization of GM foods in the state could compromise the image of the RS product in the international market: "Soybean production in RS is on the rise in the Japanese and European markets precisely because the product is natural […] We will carry out a legal analysis of the issue in order to take the necessary measures and seriously face the problem" (Hoffmann, 1999 apud Heberle, … ZH, 04/02/1999).

The opposition launched by the Dutra government against the GMOs in agriculture was justified by criticism of the "green revolution" paradigm and the affirmation of values and principles associated with an alternative family-based agricultural model linked to the agroecological paradigm. But this alternative agricultural paradigm not only implied a stimulus to productive systems, but also implied an rejection of the GM crops themselves. In its agricultural policy, there was no possibility of reconciling the agroecologically based family farming model with the GM crops. These alternatives were seen as representing incompatible agricultural models.

---

[19] In May 1998, when a strong idea emerged to establish in RS the first world niche of conventional soybeans, one of Europe's largest soymeal buyers guaranteed in Rennes, in France, that there was room for these grains. It was even suggested that the possibility that the Europe could buy the entire production of 24 entities linked to the Central de Cooperativas de Rural Producers of RS (CentralSul) (Heberle, 2005).

4.8 Back to the Beginning

Upon taking power, Dutra was faced with several challenges to implement his land reform program.[20] The main one was already given by Brazilian legislation, which indicates that only the union would have powers to carry out any policy in this area. To the states, the legislation establishes only the possibility of buying new lands, supported by laws and state funds previously created. Another possibility would be to carry out expropriations that were created for social purposes or public utility. In these cases, expropriation could only occur by compensating the owners, respecting the market value of the property. It was for this reason that the agricultural secretary of the Dutra government sought to pressure the federal government to enable the expropriation of land in the territory of the state of RS, since it was up to the federal government to make these decisions.

The agrarian law, in its article 184, assigns to the federal government the responsibility to carry out agrarian reform. This article informs that: "It is incumbent upon the Union to expropriate for social interest, for agrarian reform purposes, rural property that is not fulfilling its social function, upon prior and fair compensation in agrarian debt bonds" (Brasil, 2011). The governments that preceded the Dutra administration in the 1990s respected this precept. This means that throughout the 1990s, at least until the arrival of Olívio Dutra to the government, no other government sought to adopt any agrarian reform policy that confronted the guidelines set out in Article 184 of the Constitution. Moreover, besides following the directives of the federal policy for agrarian reform, these governments, in general, assumed the limited role that was conferred to the states in this process. The main one was to restrain illegal land invasions by means of repossession proceedings. A good example is the government of Antonio Brito that preceded the Dutra government, which, according to analyses of the period, "systematically repressed them throughout its mandate" (Ros, 2006, p. 427). The government that preceded the Dutra government in the 1990s had not, therefore, committed itself to taking any initiative to alter the guidelines for agrarian reform established by the federal government. Therefore, between 1995 and 1998, agrarian reform initiatives in the state of RS had come exclusively from the federal government. And they remained this way until the election of Olívio Dutra.

In the first half of the 1990s, land acquisitions for agrarian reform took place basically in the southern region of the state of RS. Among the reasons generally mentioned for this phenomenon is, firstly, the displacement of the MST actions to the region which, prior to this period, had directed its pressure towards the northern region. Secondly, one should consider the decrease in land prices due to their low productivity, which made the expropriation of properties in the region more likely. The northern half of the state, where the GM crops will be located some years later, therefore presented a higher agricultural yield when compared to the southern

---

[20] Except for additional considerations, the description of some main facts and events that will be reported the following can be found in the work of Ros (2006, 2009).

4.8 Back to the Beginning

region. Although the northern region had already been a region of interest to the MST, it will only become an object of interest to the movement after the release of RR soybean. It is examined below why this occurred.

The above facts may be instructive for understanding why the northern half became the MST's focus of interest at the outbreak of the conflict involving RR soybean. One should consider that the failure of MST's pressure in that region was linked to the high productivity rates of the properties located there. The accusation of illegality that the Dutra government and the MST itself will direct at the GM crops could thus change this political game, incorporating the region and the GM crops located there once again as subjects to be included in the agrarian reform process. This could be done by placing these plantations under conditions of illegality. At least this is what the MST's discourse during the conflict over the release of RR soybean will suggest, as we will see below.

In the second half of the 1990s, the agricultural sectors of the state of RS were experiencing difficulties due to the economic crisis affecting agriculture in the state. The crisis, by inducing a decrease in the agricultural productivity of farms in the region, turned many of them into fragile targets for the agrarian reform policy.[21] As a consequence, farmers of the region started to plead for a reevaluation of the productivity index criteria that were used to determine expropriations. In 1998, this context led the state of RS to witness a conflict involving INCRA's inspections in the state. The inspections were part of the agrarian reform policy that was being gradually abandoned by the Cardoso government, but which was still being promoted in various parts of the country due to the absence of an existing alternative model. The federal government will seek in 1999 to implement alternative instruments to foster agrarian reform in the region, using in the process new instruments such as auctions and negotiations. These processes would in turn be rejected by Dutra when he took over the state government because he considered expropriations to be more economical (Ros, 2006). The inspections originated in 1973 and had been created to prove compliance with the social function of the land properties. They allowed to gauge the rates of agricultural productivity becoming the main instrument of expropriation of land for the government.[22] On the other hand, although the inspections were still carried out for this purpose, the federal government was already showing signs of wanting to incorporate market instruments such as auctions and land purchases. In this new policy, the inspections would still play

---

[21] The crisis manifested itself through loss of profitability in agriculture and cattle ranching the reduction of areas planted in crops and the concentration of grain production in certain establishments. One consequence of this process was the decrease in the price of land and the increase in offers of rural properties to INCRA. On these points, see Ros (2009, p. 14).

[22] Between 1995 and 1998, initiatives to promote agrarian reform were taken exclusively by the federal government. In this phase, the state government of Antônio Britto (PMDB) limited itself to demanding the eviction of land when it was occupied by the MST.

an important role in fostering land reform, but their function would be distinct from that exercised until then.[23]

While the MST still pressured the federal government to maintain an agrarian policy based on expropriations, the large rural farmers were increasingly dissatisfied with it. This served to intensify the conflict in the region. Their main criticism was the outdated productivity indices that were used to support the inspections. For this reason, the reevaluation of these indices was one of the main points in the struggle of the large landowners during the conflict. For them, the inspections were perceived as flawed processes that led to a lack of independence of the agencies responsible for land reform. This perception suggested, in the first place, an alignment of officials and representatives of government agencies, especially INCRA, with the causes of rural social movements. Second, this bias was seen as associated with the outdated basis of calculation used to measure the agricultural productivity of farms in the state (Ros, 2009, p. 251). The dissatisfaction of farmers was also fed by the permanent and growing pressure exerted by the MST, which intensified its actions in favor of land expropriation in the midst of the economic crisis. Finally, with Dutra's election, farmers found themselves unprotected due to the absence of a government to guide such conflicts. This situation led farmers to organize themselves more and more in order to oppose the actions of the MST in the state and the agrarian policy guidelines of the state and federal governments.

In 1998, the INCRA superintendency in RS tried to restart the re-registration of rural properties in the municipality of Bagé. In response, large landowners began to carry out a series of actions that took the form of what became known as the "Vistoria Zero" movement. The actions included paralyzing government inspections and blocking roads in order to prevent inspectors from entering the properties. The objective of these actions was to make it impossible to verify productivity indexes, thus preventing the process that would lead to productivity evaluations and possible expropriations.[24] The argument presented by Farsul and the large farmers was the need to review the productivity indices that guided the inspections. The success of the action would serve as a reference for similar situations and it was from that moment on that the federal government would seek to stimulate the use of alternative instruments to foster agrarian reform in the state. Thus, as Ros (2006, p. 229) points out, the "Vistoria Zero" movement created a pattern of political action that would always be activated at times involving the occupations and the actions of INCRA.

---

[23] Whereas the federal government had already been considering changing the expropriation process, as it had been carried out, its permanence in the agrarian reform policy in RS was seen as a result of pressures exerted by the MST itself in the region. For the case of the state of RS, it is suggested that the "advance of the inspections is directly related to INCRA's need in responding to the political pressures unleashed by the MST" (Ros, 2006, p. 227).

[24] The impact of the action of the large farmers in this period was so significant that Gedeão Pereira Silveira, then representing Farsul, describes its effects as follows: "From that moment, the agrarian reform practically did not avenge in the state of RS through the rural property expropriation" (Silveira, 2004 apud Ros, 2006, p. 26).

4.8 Back to the Beginning

With the actions developed in the "Vistoria Zero" movement, farmers from the state of RS began to perceive their capacity to influence the land reform policy promoted by the state until then. The barriers at the entrance to rural properties can be seen as comparable, as Ros (2009) notes, to the MST's own occupations, since they also constituted a type of direct action promoted by large agricultural producers. In doing so, they also developed a sense of unity of collective action among large agricultural producers.

With the election of Olívio Dutra, the federal government will sign a Technical Cooperation Agreement with the state government to resume agrarian reform in the region and, with it, the property inspections. The general lines of the agreement will be established at a first hearing, at which the Dutra government representative will express special interest in taking over the inspections. To this end, he will invoke Provisional Measure No. 1.703 of November 1998. It was stated in Art. 3 of that law that:

> Art. 3 The Union, through an agreement, may delegate to the States the registration, surveys and assessments of rural properties located in its territory, as well as other attributions related to the execution of the National Agrarian Reform Program, observing the parameters and criteria established in federal laws and normative acts. (Brasil, 2011)

Law No. 1703 allowed the states to carry out inspections with the aim of promoting agrarian reform. But this joint work presumed a kind of division of labor between the state and federal governments. Thus, activities such as property registration and productivity inspections were delegated to the local state government. However, as can be seen in this article, this could only occur once "the parameters and criteria established in federal laws and normative acts" (Brasil, 2011) were observed. In short, although the inspections could be carried out by the state government, they should be subject to the guidelines and legal bases of the federal government's own land reform policy. The federal government responded positively to the Dutra government's proposal, in which it would assume the evaluations and inspections of the properties. Once the agreement was established, the federal government would then request the creation of a working group to examine the issue.

This point is of particular importance for what will be examined below on the tensions involving GMOs in the region. This is because, while the state government's implementation of the inspections could be supported by Law No. 1703, many of the elements of the agrarian reform policy that Dutra intended to implement in RS clashed with the same guidelines of federal agrarian reform policy. This suggests, a difference in interpretation and interests that each government (state and federal) was offering to the process. The Dutra government saw the inspections as having a role quite distinct from that of the federal government, although the former depended on the consent of the latter.

An essential element to keep in mind when analyzing the realization of this agreement is the distinct visions that shaped the agrarian reform policies of the federal (FHC) and state (Olívio Dutra) governments. Both governments will celebrate the agreement based on different, if not contradictory, guidelines for agrarian reform. For the federal government, the resuming of inspections would be done, as shown before, through a new model of agrarian policy. In this model, the inspections would no longer be linked to forced legal expropriation without any financial compensation for the owners, but would be induced by means of auctions or by acquisition of the properties through the financial resources from the state itself (Banco da Terra). The Dutra government would enter into the agreement expressing a distinct and even opposite view. The state government became responsible for surveying the properties without assuming the new guidelines of the federal government. This meant that the state government had greater power to carry out the inspections, while at the same time, according to government representatives, continuing the policy of expropriation. As the Secretary of Agriculture of the Dutra government stated during the period of the agreement:

> We excluded the Land Bank, we excluded I don't know how many other things. We made an agreement that added up where we had an interest in adding up, which in this case was to accelerate expropriations, the use of agrarian debt bonds, and that the money from the state treasury would go exactly where INCRA had the most limitations, which is the issue of improvements, productive infrastructure, and so on. We were going to make a symbiosis, but we were not going to reinforce the model conceived by them of decentralization, within that new concept of the rural that was, in our opinion, very complicated. (Hoffmann apud Ros, 2006, p. 241)

In the same period, José H. Hoffmann will also declare that: "As far as it depends on us, one hundred percent of the settlements will be made through expropriations, much more economical than purchases" (Hoffmann apud Ros, 2006, p. 242). This explains the Dutra government's decision to exclude the Banco da Terra from the land reform process. Thus, the decentralization proposed by the federal government, by offering greater power and autonomy to the state in the agrarian reform process, was seen as an ally for Dutra to pursue his own agrarian policy. But a policy quite distinct from the one that the federal government was envisioning for the region.

The agreement was received with great distrust and fear by the state's farmers. They feared that the agreement could be used to return to the traditional agrarian policy, in which expropriations would result from a strictly legal process. And this would occur with three aggravating factors. First, this would be done through the use of state legal inspections that would make use of outdated agricultural productivity indexes. They would be conducted by a leftist government aligned to a social movement such as the MST, which was part of the government's administrative structure. And, third, they would be conducted by a government that affirmed that it would not follow the federal government's guidelines. As seen in the previous part, the state secretary of agriculture himself, José H. Hoffmann, went so far as to affirm that the agreement would make it possible to accelerate agrarian reform through expropriations that, according to him, were "much more economical." The Dutra government itself made it a condition for the state government to participate in the agreement signed with the federal government.[25] As a result, the farmers would strongly criticize the agreement, as it was "designed to strengthen the instrument of

---

[25] For more details on this part of the conflict, see Ros (2006).

4.8 Back to the Beginning

expropriation, both on the part of INCRA and the state government" (Ros, 2006, p. 242). This indicated that, at least for large farmers, the inspections would be resumed without revision of agricultural productivity indices, in a context of economic crisis and conducted by a government that had the MST in its Agriculture Secretariat.[26] Considering the context through which the new agreement was being announced by the state government, it is understandable that Farsul, the main entity representing large farmers, rejected the agreement entirely.[27] It is likely that the large landowners also felt increasingly isolated and without the same legal and political guarantees they enjoyed until that moment.

Although the agreement was planned to promote an agrarian reform, all the elements for the conflict to intensify were given. In an ad in the newspaper, Correio do Povo, the legal advisor of Farsul, the main representative entity of large agricultural producers, warned about the "risk of having again serious turbulence in the state," because, according to him, "we cannot play with the producer" (Correio Do Povo, apud Ros, 2006, p. 242). For him, the protest against the agreement occurred because "the Secretary of Agriculture himself, José H. Hoffmann showed that 'he has a side, and his side is the Landless Workers' Movement'" (Correio Do Povo apud Ros, 2006, p. 242). The legal representative of Farsul would also state that the main person responsible for agrarian reform in the Dutra government, Friar Sérgio Görgen, presented himself as a "notorious land invader." Gedeão Pereira Silveira, then president of Farsul's Land Issues Commission, following the same criticism, would say that the agreement represented a situation with the "MST doing inspections within the state" (Correio Do Povo apud Ros, 2006, p. 242). The climate that had been established in local politics was therefore one of total distrust between farmers and government.

The position of the federal government somehow contributed to disseminating this tension. On the one hand, Raul Jugmann, then Minister of Agriculture, sought, through various pronouncements to calm the large farmers, warning that the inspections were suspended. On the other hand, the relationship between members of INCRA, a federal government agency, and the MST seemed to say something different (Ros, 2006). In this context, the state government did not appear to be a guarantor of the legal order that could ensure property rights for farmers. And on the other hand, the Dutra government seemed to create the conditions for the MST's own pressures to now become much more effective. The main factor influencing this process was his government's willingness to give political treatment to land

---

[26] Farmers' actions against the inspections in 1997 were directed against INCRA and, therefore, signaled a conflict between ranchers and a federal government agency. With the construction of the covenant, the conflict that will now be drawn will present a more regional scale, since the actions of resistance against the state legal inspections will be directed against the state government itself, which, as from the agreement, will be the main party responsible for carrying out the process of monitoring of agricultural productivity on farms.

[27] In the period, Carlos Sperotto (1999 apud Ros, 2006, p. 241), president of Farsul, will state that: "We find the position of resumption of the inspections strange, without revised productivity rates. It is a totally untimely manifestation."

invasions. Thus, the Dutra government would claim that land occupations would cease to be a "police case" and become a "political case." With these words, he suggested that his government, unlike the previous one, would no longer take the measures usually applied to promote land evictions. If this were to occur, they should be submitted to a process of negotiation.[28]

The crisis facing agriculture in the state of RS during this period posed several problems for agrarian reform measures. The government's actions in this area required the measurement of agricultural productivity levels on farms. With the crisis, this led farmers in the state to call for a reassessment of the criteria for agricultural productivity. The discussion and the attempt by farmers from Rio Grande do Sul to review these criteria permeated various moments of the conflict, since it was through these criteria that INCRA, and later the state government itself, could justify the expropriation of land. Let us consider that these tensions emerged at the end of 1998, when CTNBio decided to release RR soybeans and when several news items involving the irregular planting of RR soybeans started appearing in the regional media. Thus, the conflict involving the inspections reappeared at the same time that RR soybean was being released by the federal government. How did these two episodes end up intertwining then?

4.9 GM Soybean: A Friend of Agrarian Reform?

In examining the conflict in the region, some works suggest, as is the case of Menasche (2002), the existence of a process of transvaluation that would indicate an "assimilation of particular circumstances to a broader, collective, lasting cause or interest, and therefore less dependent on contextual conditions" (Tambiah, 1997). The "particular circumstances" would, in this case, involve the various disputes involving the GM crops, while the "broader cause or interest" would be associated with the agrarian reform process in the region. This transvaluation could be verified when one realizes that the inspection action of clandestine GM crops, one action among others developed by the Dutra government to bar GM crops in the region, became associated with the "application of the threat of land expropriation for Agrarian Reform" (Menasche, 2002, p. 234).

In short, the actions that the Dutra government claimed were aimed at supervising the planting and commercialization of GM seeds in the south of the country

---

[28] As the following research examining the conflict will indicate: "The sign that in its mandate [...] land conflicts would be treated as 'political' rather than 'police' cases created a favourable environment for increased social pressure via land occupations." This also means that if the properties were invaded, the invaders would not be removed from the properties by police forces. The process should take place through some form of "negotiation" (Ros, 2009, p. 38).

4.9 GM Soybean: A Friend of Agrarian Reform?

were interpreted as having another objective by the large farmers.[29] And by examining the previous history, one can see some reasons for this to have occurred. This can be confirmed by the very perception that these actors expressed about the actions involving the inspections of properties in the region. The perception that came to predominate among pro-Tg farmers was precisely the view that the latter had become an obstacle to the Dutra government's agrarian policy and that the Dutra government's surveillance of clandestine GM crops, an action taken by the Dutra government in the period, would become a threat to the expropriation of land in order to make the state government's agrarian reform policy viable. This perception becomes visible when a representative of the pro-GMO farmers suggests that the GM crop issue was being used by the Dutra government as a "façade" so that large farmers' properties could be "expropriated for land reform purposes, for the MST [...]" (Leader of the Clube Amigos da Terra *apud* Menasche, 2002, p. 233). Another leader, along the same lines, argued that the Dutra government "was not concerned with the issue of transgenics, but with other issues" (Leader of the Clube Amigos da Terra *apud* Menasche, 2002, p. 234). These issues would be associated, according to the latter, to the fact that the "state government, in its ideology," does not accept "agriculture other than family agriculture" (Leader of the Clube Amigos da Terra *apud* Menasche, 2002, p. 234). This view was also presented by Monsanto's own representative, Luiz A. do Val, the company's regulatory president. On March 9, 1999, he stated that the ban on GM crops in the state was presented as "an ideological measure" since the "gaúcho government (of RS) is left-wing and linked to nongovernmental organizations that associate biotechnology with the power of large multinational companies".

These links demonstrate the factors that have fueled the distrust of pro-Tg farmers in the conflict, a distrust that was stimulated by the MST's own discourse in the dispute. The organization itself went so far as to state that the areas planted with RR soybeans should be set aside for agrarian reform, while requesting that the Public Ministry carry out a rigorous investigation into the stimulus and inducement to plant RR soybeans in the state (Menasche, 2002). The "lands cultivated with transgenics have to be framed" declared an MST leader of the period, "in the same law that represses psychotropic plantations" (ZERO HORA apud Pelaez & Schmidt, 2000, p. 27). MST members not only declared that the GM crops should be earmarked for agrarian reform, but also asked the Public Ministry for a rigorous investigation into the plantations.

The psychotropic law provides that, in the case of irregular plantations, agricultural properties are liable to expropriation. What is striking about the MST leader's words is not the equivalence made between psychotropic plantations and GM crops,

---

[29] This process should not be considered extraneous to environmental policy analysis. One of the precepts of international ecopolitics proposed by Le Preste, for example, suggests that the "environmental protection is not only an end in itself. It is equally a means to achieve other policy objectives, such as the democratization of public life or the decentralization of power," which makes him conclude that "environmental problem is rarely a simple environmental problem" (Le Preste, 2001, p. 31).

but his proposal that, within this type of legal framework, the possibility of expropriating the land on which the GM crops were located should be opened up, a clear suggestion that GM crops could be the focus of agrarian reform policies, as the large rural producers had suspected. And this statement was made by an entity that was part of the government's hard core in the agricultural area. If all this were true and had a legal basis, then the inspections of GM crops could serve to produce the expropriations of farmers' lands that MST members were targeting. The fear of expropriation by pro-Tg farmers was therefore backed up by the very words of MST leaders who, until that point, were also seen as spokespersons for the government. Thus, the MST's discourse, linking possible land expropriations to the inspection of GM crops, was interpreted as a strategy to make agrarian reform viable in the state. For the pro-Tg farmers, these approximations made the MST's position and that of the government inextricably linked, especially since some members of the movement were part of the government at the time. At the same time, neither the governor nor the Secretary of Agriculture, José H. Hoffmann, denied any such intention at the time. This situation could possibly confirm that the pro-Tg farmers were consenting to what the MST was proposing.

It should be added here the position of the Dutra government itself, which, in a way, will give even more credibility to the words of the MST members. The Secretary of Agriculture himself, commenting on the project that sought to bar the commercialization of RR soybean in the state, will state that:

> What we have done until now is to apply Law n° 9.453, which makes a series of requirements for the realization of these experiments that, for now, have not been met. As these are met, obviously from the legal point of view, there will be no direct and objective prohibition, except those arising from environmental laws and the seed law, which will be applied with equal rigor. (Hoffmann, 1999, p. 174)

It should be considered that the state had no concrete policy to back up the laws that the Dutra government sought to defend in order to supervise GM crops. So much so that on November 24 and 25, 1999, at an International Seminar held in Brasília, Hoffmann will declare that:

> Our project for the free territory covers an inspection structure. In Rio Grande do Sul, there is the Serviço de Classificação de Produtos de Origem Vegetal (Service for Classification of Vegetable Origin Products), structured in 54 municipalities, especially on the border, where products such as rice, soybeans and corn are classified. This structure will be adapted and expanded to include inspection of soybeans, including those coming from other countries. (Hoffmann, 1999, p. 174)

This declaration was made in June 1999. In that same period, an agreement had been signed between the state and federal governments to promote land reform in the state. If the government were to consider GM crops illegal, as its discourse suggested, would it apply the seed law that would allow them to be classified as illegal plantations? What implications would this have for pro-Tg farmers in the context of an ongoing land reform process? And with members of the government, linked to the MST, who claimed that GM crops would be legally framed in the same way as psychotropic plantations. In this scenario, more like an enemy, it is not impossible

4.9 GM Soybean: A Friend of Agrarian Reform?

to imagine here that the RR soybean were considered allies for the policy that the Dutra government and the MST sought to implement in the state. This is because the GM crops could open up new possibilities for agrarian reform policy in the region. At least what is clear from this discourse is that the GM crops were being used as a threat of expropriation.

Among other things, if the government were to use such an expedient, such actions would easily allow low levels of productivity to be attributed to properties where GM crops were classified as illegal. Under this condition, productivity could even be considered zero, thus setting a precedent for land expropriation in the region. Moreover, such a strategy would redirect land reform to the northern region, where the soil is more propitious for planting, thus satisfying a long-standing demand by the MST for this land or part of it. It would make it possible to speed up agrarian reform, since in the Agriculture Secretary's view, expropriation was the quickest and most efficient way to promote this process. The government would not only placate the large landowners, but would also satisfy the demands of the sectors that elected Olívio Dutra, winning admirers both inside and outside Brazil, with a government that would implement an agrarian reform based on agroecological family farming. But it is difficult to say to what extent some of these factors were part of the political calculation of those who were in favor of including GM crops in the agrarian reform process. In any case, the fact that GM crops were seen as a target for agrarian reform policy indicates that, in some way, some of these possibilities may have been considered by political actors allied to the Dutra government. After all, there must have been some reason for the MST to threaten GM crops with their inclusion in the land reform policy. If not, why would it threaten to expropriate them in this context?

The tensions brought about by the commercial release of RR soybean involved a geographical change in the conflicts, which throughout the 1990s had been concentrated in the southern region of the state. This had occurred for the reasons examined above: decrease in land prices, changes in the MST's strategy, etc. As the government directed its expropriation policy towards the southern region, the MST then directed its actions towards the same region in order to advance the land expropriation process through popular pressure. However, with the release of the RR soybean, the focus of the conflict was redirected once more to the northern region of the state, where, until then, the MST itself had found it difficult to exert pressure for its political agenda. In this context, the possibility of classifying the GM crops as illegal in the northern part of the state indicated the possibility of redimensioning the pressure strategies in this region, which, as already mentioned, presented a higher quality of soil for planting. This could occur to the extent that the plantations were seen as illegal, annulling their production for the purposes of assessing productivity. Thus, the MST's strategy could possibly rely on the fact that the most productive regions could now be expropriated for land reform.

The interpretation given by pro-Tg farmers to the government's intentions corresponds, in a way, to the very discourse that it, the Dutra government, presented to public opinion in the period. First, the inspections of agricultural properties by the political organs of the State of RS were, throughout the 1990s, associated with the agrarian reform policy in the region. Their objective was to recognize unproductive areas in order to include them in the agrarian reform policy. In other words, if these lands were classified as unproductive, they could be targeted, at least in principle, for expropriation. The Dutra government not only failed to clarify the possible meaning of these inspections, but also confirmed the farmers' fears when members of the MST, who were part of the government, boasted about the threat of expropriation of land where GM crops were located. Because it would be possible to classify GM crops as illegal, this would make it possible to subsequently define them as unproductive. This in turn would nullify their productivity rates. Moreover, what further fueled this mistrust was the fact that the policing of GM crops, in the perception of pro-Tg farmers, was a function of the federal government. If this was true, as the court decision subsequently confirmed, what were the purposes of the Dutra government's inspections? These doubts therefore reinforced the pro-Tg farmers' perception that the Dutra government's inspection process could conceal ulterior motives. This suspicion was reinforced by the fact that Dutra's agrarian reform policy was clearly different from that of the federal government.[30]

This whole situation triggered a stronger movement involving pro-Tg farmers in November 1999. From that moment on, many of them began to defend not only GM crops, but also the right to property. The two agendas became integrated. Hence, at some points during the conflict, some of them no longer raised banners aimed only at defending the commercialization of RR soybean, but would call for "military force to protect property" (Bauer, 2006, p. 226). It is clear, therefore, that, for these farmers, the objective of inspecting the GM crops concealed an objective of expropriation of agricultural properties. It was for this reason that the most tense actions of the conflict in the region began to emerge in this period, which in turn would reproduce the logic of resistance that had begun with the "Vistoria Zero" movement examined earlier. The same actions that the large farmers had taken in 1998 with this movement were then reproduced to prevent the inspections on the GM crops. This was not only because it was possible to reproduce a political action that had proved effective in the past, but also because the farmers were threatened by the same problems. The government's discourse of monitoring the GM crops, as noted earlier, was seen as a "façade" by pro-Tg farmers who began to resist the government's actions. These actions were now being reproduced to invalidate the inspections, the purpose of which was not properly clarified at the time.

---

[30] Let us recall here the announcement of the Secretary of Agriculture, Jose Hoffmann, who informed that the agrarian reform would be carried out by "expropriations, much more economical than purchases." (Hoffmann, 1999 *apud* Ros, 2006, p. 242).

4.10 Final Considerations

The Dutra government faced two major challenges in its campaign against the commercial release of RR soybean. Its policy on land reform clashed with the federal government's guidelines and directives in this area. And the same had occurred with his policy on GMOs, which, in a way, was linked to his policy on agriculture. Unlike the Dutra government, both CTNBio and the federal government had a very distinct view on the two issues and on the possible links between them. Agrarian reform was never considered by CTNBio as an issue related to the commercial release of GMOs in agriculture. The view that RR soybean could only bring benefits to farmers, expressed in Bresser-Pereira's considerations on the unfeasibility of making a decision founded on a trade-off, exempted the commission from trying to identify any distributional impact on small farmers, which made it see the commercial release of RR soybean as a kind of positive game in which all players in the political process would win.

If at the federal level, the Dutra government was faced with this conflict, internally, the disputes developed without the government having a solid political base to carry out its political agenda. Most of the legal actions that Olívio Dutra tried to implement against the commercial release of the RR soybean failed due to the little political support that the government had managed to mobilize among political leaders within the state itself. The laws that the governor sought to institute to implement his plan for an "Um Território Livre de Transgênicos" were simply not approved by the state legislature. And, even if they did win any approval, it must be considered that they would be in direct confrontation with federal guidelines and laws. This could produce a legal confrontation with little chance of victory for Dutra.

It is understandable that, in the face of the attempt to create an "Um Território Livre de Transgênicos," the government clamored for the need to obtain greater political decision-making autonomy. After all, its policy of resistance to GMOs would only be successful if the state had greater political autonomy to deliberate on the issue. But this also required greater autonomy to deliberate on the process of agrarian reform in the state. Which, again, was something it lacked. In addition, the governor's own party took a pro-liberation stance for the GMOs at the federal level later on (Lula government), following an orientation opposite to that presented by the governor's party in the RR soybean liberation period. For all these reasons, the "Um Território Livre de Transgênicos" campaign seemed to have little chance of succeeding if one considers the objectives that the campaign had set for itself. On the other hand, the resistance to RR soybean in the south of the country helped to amplify the debate on GMOs in Brazil and the world. It has, in a way, helped to amplify the debate for regions and areas of national policy for the GMOs. It would not be wrong to say that the debates to be examined in the following chapters were also influenced in some way by it.

References

Acselrad, H. (2010). Ambientalização das lutas sociais: o caso do movimento por justiça ambiental. *Estudos Avançados, 24*(68), 103–119.
Almeida, J. (1999). *A construção social de uma nova agricultura*. EdUFRGS.
Almeida, J., & da Silveira, C. A. (2000). Significados sociais das biotecnologias: o campo de disputas em torno das sementes transgênicas no Rio Grande do Sul. Reunião Anual da Anpocs, 24., 2000, Anais [...]. Petrópolis.
Bauer, M. W. (2006). Garantindo os benefícios de uma moratória. Lavouras transgênicas no Brasil – especialmente a de soja (1995–2004). Disponível em: https://www.researchgate.net/publication/30524626_Garantindo_os_benefi-cios_de_uma_moratoria_Lavouras_transgenicas_no_Brasil_especialmente_a_de_soja_1995_-_2004. Acesso em: 10 ago. 2020.
Brasil. (1988). *ART. 184: Dispõe sobre a Política Agrícola e Fundiária e da Reforma Agrária*. Disponível em: http://www.planalto.gov.br/ccivil_03/Constituicao/Constituicao.htm. Acesso em: 10 jul. 2011.
Brasil. (1997). *Lei n° 9456, de 25 de abril de 1997. Institui a Lei de Proteção dos Cultivares e dá outras providências*. Diário Oficial da União. Disponível em: www.inpi.gov.br. Acesso em: 5 set. 2005.
Brasil. (2007). *Projeto de decreto legislativo no. 90, de 2007. Susta a aplicação do artigo 3° do Decreto no. 4.680, de 24 de abril de 2003, que regulamenta o direito à informação, assegurado pela Lei no. 8.78 de 11 de setembro de 1990*. Diário do senado federal.
Bresser-Pereira, L. C. (1999). Alimentos transgênicos e biossegurança. Disponível em: http://www.bresserpereira.org.br/view.asp?cod=596. Acesso em: 19 ago. 2020.
Bryant, B. (1991). Introduction. In B. Bryant (Ed.), *Environmental Justice. Issues, policies and solutions*. Island Press.
Bullard, R. (2004). Enfrentando o racismo ambiental no século XXI. In H. Acselrald et al. (Eds.), *Justiça ambiental e cidadania*. Relume Dumará.
Doyal, L., & Gough, I. (1991). *A theory of human need*. Macmillan.
Fernandes, G. B. (2010). *Os Direitos dos Agricultores no Contexto do Tratado de Recursos Ditogenéticos da FAO – O Debate no Brasil. Documento produzido pela AS-PTA*. Disponível em: http://aspta.org.br/files/2011/05/Os-direi-tos-dos-agricultores-no-contexto-do-tratado-de-Recursos-Fitogen%C3%A9ti-cos-da-FAO.pdf. Acesso em: 6 ago. 2020.
Figueroa, R., & Mills, C. (2003). Justiça ambiental. In D. Jamieson (Ed.), *Manual de Filosofia Ambiental*. Instituto Piaget.
Gerchmann, L. (2003). *MST planta soja transgência no Rio Grande do Sul*. Folha de São Paulo. Disponível em: https://www1.folha.uol.com.br/folha/dinheiro/ult91u64506.shtml. Acesso em: 22 mar. 2012.
Gonzales, C. G. (2007). Genetically modified organisms and justice: The international environmental justice implications of biotechnology. *Georgetown International Environmental Law Review (GIELR), 19*, 583. Disponível em: http://papers.ssrn.com/sol3/papers.cfm?abstract_id=986864. Acesso em: 26 ago. 2008
Gould, K. A. (2004). Classe social, justiça ambiental e conflito político. In H. Acselrald et al. (Eds.), *Justiça ambiental e cidadania*. Relume Dumará.
Haland, W. (1999). On needs: A central concept in the Brundtland report. In W. M. Lafferty & O. Langhelle (Eds.), *Towards sustainable development*. Macmillan.
Heberle, A. L. O. (2005). *Significações dos transgênicos na mídia do Rio Grande do Sul*. Tese (Doutorado em Ciências da Comunicação) – Universidade do Vale do Rio dos Sinos. Programa de Pós-Graduação em Ciências da Comunicação.
Herculano, S. (2002). Riscos e desigualdade social: a temática da justiça ambiental e sua construção no brasil. In: *I Encontro da ANPPAS* (pp. 1–22). ANPPAS. Disponível em: https://www.professores.uff.br/seleneherculano/wpcontent/uploads/sites/149/2017/09/Riscos__v4_e_desigualdade_social.pdf. Acesso em: 10 maio 2021.

References

Hoffmann, J. H. (1999). Fundamentos técnicos e jurídicos para a instituição de áreas livres de transgênicos. In: *Seminário Internacional Sobre Biodiversidades, 1999, Brasília*. Anais [...]. Senado Federal.
Langhelle, O. (1999). Sustainable development: Exploring the ethics of our common future. *International Political Science Review, 20*(2), 129–149.
Le Preste, P. (2001). *Ecopolítica internacional*. Senac.
Leff, H. (2002). Agroecologia e saber ambiental. *Agroecologia e Desenvolvimento Sustentável, Porto Alegre, 3*(1).
Lenzi, C. L. (2022). *Environmental sociology: Risk and sustainability in modernity*. Lexington Books.
Menasche, R. (2002). Legalidade, legitimidade e lavouras transgênicas clandestinas. In H. Alimonda (Ed.), *Ecologia política: naturaleza, sociedad y utopia*. CLACSO.
MST. (2006). *Transgênicos: a serviço de quem?* Disponível em: http://www.mst.org.br/node/10130. Acesso em: 28 jul. 2010.
Pelaez, V., & Albergoni, L. (2004). Barreiras técnicas comerciais aos transgênicos no Brasil: a regulação nos estados do sul. *FEE Indicadores Econômicos, Porto Alegre, 32*(3), 201–230.
Pelaez, V., & Schmidt, W. (2000). A difusão dos OGM no Brasil: imposição e resistências. *Estudos Sociedade e Agricultura, 14*, 5–31.
RBJA. Rede Brasileira de Justiça Ambiental (Brasil). (2010). *Princípios da Justiça Ambiental*. Disponível em: http://www.justicaambiental.org.br/_justicaambiental/pagina.php?id=229. Acesso em: 16 mar. 2011.
Rede De Agricultura Sustentável. (2000). *Transgênicos: uma cronologia*. Disponível em: http://www.agrisustentavel.com/trans/crono.htm. Acesso em: 27 jul. 2012.
Rio Grande do Sul. (2014a). *Assembleia Legislativa. Proposição: PL 16 1999. Veda o cultivo comercial de organismos geneticamente modificados (OGMs) no Estado do Rio Grande do Sul e dá outras providências*. Disponível em: http://www.al.rs.gov.br/legislativo/ExibeProposicao/tabid/325/SiglaTipo/PL/NroProposicao/16/AnoProposicao/1999/Origem/Px/Default.aspx. Acesso em: fev. 2014.
Rio Grande do Sul. (2014b). *Assembleia Legislativa. Decreto n° 39.314, de 3 de março de 1999. Regulamenta a Lei n° 9.453, de 10 de dezembro de 1991, que dispõe sobre pesquisas, testes, experiências ou atividades nas áreas da Biotecnologia e da Engenharia Genética, e dá outras providências*. Disponível em: http://www.sema.rs.gov.br/sema/html/dec_39314.htm. Acesso em: fev. 2014.
Rio Grande do Sul. (2014c). *Assembleia Legislativa. Lei n° 9.453, de 10 de dezembro de 1991. Dispõe sobre pesquisas, testes, experiências ou atividades nas áreas da Biotecnologia e da Engenharia Genética, e dá outras providências*. Disponível em: http://www.mp.rs.gov.br/hmpage/homepage2.nsf/pages/cma_dec39.314. Acesso em: fev. 2014.
Rocha, G. (2012). *Crítico da transgenia, MST planta soja com alteração genética em assentamento*. Disponível em: http://www1.folha.uol.com.br/folha/brasil/ult96u493491.shtml. Acesso em: 18 ago. 2012.
Ros, C. A. D. (2006). *As políticas agrárias durante o governo Olívio Dutra e os embates sociais em torno da questão agrária gaúcha (1999–2002)*. Tese (Doutorado em Ciências Humanas e Sociais) – Universidade federal Rural do Rio de Janeiro, Instituto de Ciências Humanas e Sociais.
Ros, C. A. D. (2009). O movimento "vistoria zero" e a resistência do patronato rural às políticas de assentamentos no Rio Grande do Sul. *Sociologias, Porto Alegre, 11*(22), 232–278.
Schlosberg, D. (2007). *Defining environmental justice: Theories, movements, and nature*. Oxford University Press.
Seifert, F., & Torgersen, H. (1997). How to keep out what we don't want: An assessment of 'Sozialverträglichkeit' under the Austrian Genetic Engineering Act. *Public Understanding of Science, 6*, 301–327.
Shrader-Frechette, K. (2002). *Environmental justice. Creating equality, reclaiming democracy*. Oxford University Press.

Tambiah, S. J. (1997). Conflito etnonacionalista e violência coletiva no sul da Ásia. *Revista Brasileira de Ciências Sociais, São Paulo, 12*(34).

Thompson, P. B. (1997). *Food biotechnology in ethical perspective*. Champman & Hall.

Veras, M. M. (2005). *Agroecologia em assentamentos do MST no Rio Grande do Sul: entre as virtudes do discurso e os desafios da prática*. Dissertação (Mestrado em Agroecossistemas) – Universidade Federal de Santa Catarina.

Wilkinson, J. (2004). *Distintos enfoques e debates sobre a produção familiar no meio rural*. Disponível em: www.emater.tche.br/docs/agroeco/revista/n3/06-ar-tigo1.htm. Acesso em: 6 ago. 2020.

Chapter 5
Science in Dispute: Sound Science and the Conflict over Risk Analysis

> *As a British scientist told me: in this case, it is not the genetically modified foods that are at stake, it is science. (Luiz Carlos Bresser-Pereira – Minister of MS&T)*
>
> *The Member argued that we were fighting science against politics. This argument is false, it is not science against politics, but science against science. (Deputy Fernando Gabeira)*

In this chapter, the second axis of the conflict associated with the commercial release of RR soybeans will be examined. This dimension of the conflict will be referred to here as the uncertainty story line. This part of the dispute brings us back to issues of environmental risk posed by the commercial release of RR soybean and issues of scientific uncertainty that permeated the decision-making process. The story line of uncertainty is formed by the discursive conflict between two distinct political alliances. The liberation alliance is represented especially by CTNBio, MS&T, and other sectors of the scientific, political, and economic fields. This political alliance is characterized by a minimal approach to PP. The preventive measures that are seen as necessary for RR soybean liberation tend to be reduced to the realization of an RA in this discourse. However, as will be seen below, it has a number of assumptions about the idea of precaution and issues related to risk perception and scientific uncertainty that distinguish it from what can be termed a precautionary political alliance. The latter is represented by organizations such as IDEC and Greenpeace and other sectors of society that have offered it support. This discourse is characterized precisely by a contestation of the RA as a sufficient and satisfactory instrument in the policy-making process that underpinned the commercial release of RR soybean. In the case of the story line of uncertainty, the political alliance of precaution is the main actor in its structuring. From the analysis of this conflict, it is possible to verify two distinct frames of precaution that are associated with the discourses of these groups. These elements are summarized in Table 5.1. In the remainder of this chapter, these structuring points of the uncertainty history line will be analyzed.

Table 5.1 The dispute about risk and precaution

Structuring issue *(issue framing)*	Liberation political alliance	Precaution political alliance
PP in general	Precaution as RA. Preventive measures are restricted to RA and are reduced to techno-scientific procedures.	Precaution beyond RA. Associated with the conduct of environmental impact assessments (EIAs). Precaution involves both scientific and normative elements.
Triggering the precautionary principle	RA as a sufficient condition for the application of preventive measures. Scientific uncertainty as nonproblematic for decision-making.	RA as insufficient to implement preventive measures. Recognition of scientific uncertainty as a condition for application from the PP.
RA, EIA, and the precautionary principle	RA and EIAs are seen as similar or as equivalent methodologies for assessing the impacts of RR soybean.	EIAs are more scientific and reliable than RA. RA and EIA are different impact assessment methods, and RA is an insufficient tool for assessing the risks associated with RR soybean.
Uncertainty and state of knowledge	The knowledge is sufficient and reliable to allow the commercial release of RR soybean. There is a "consensus" and "scientific proof" of GM safety. Scientific uncertainty is seen as "normal," "speculative," or "nonexistent."	There is no certain knowledge about the adverse effects of RR soybeans to commercial release it for trade. Scientific uncertainty does not allow the existence of safety for RR soybean to be asserted. Uncertainty can be reduced through scientific research ("need to know more").
Regulation	Risks can be managed through conventional means or means that require little change to the regulatory framework.	The commercial release of RR soybean requires stronger regulatory measures, involving the creation of more knowledge, more public participation, and more enforcement.
Evidence of danger (guideline)	RR soybean can be considered as safe as conventional soybean. Use of the principle of substantial equivalence.	Scientific research does not confirm the safety of RR soybean. The principle of substantial equivalence is a pseudoscientific concept.
Cost/benefit analysis	The decision on the commercial viability of soybean is up to the market. It is the economic agents who determine whether or not they want to use it. Individual preferences should be delegated to the market. The absence of ecological and human health risk creates the conditions for the commercialization of RR soy.	The commercial viability of soybean cannot be left entirely to the market, as its use may have implications for those who wish to opt for traditional agriculture. The uncertain picture surrounding the risks of RR soy requires stricter regulatory measures.

(continued)

Table 5.1 (continued)

Structuring issue (issue framing)	Liberation political alliance	Precaution political alliance
Burden of proof	The absence of information on adverse effects attests to the safety of soybean. The scientific consensus and the principle of substantial equivalence shift to critics the need for proof of harmfulness of RR soybean.	RA does not reduce scientific uncertainty. The EP and AR, as risk assessment tools, are insufficient and do not generate credibility for the policy decision. The burden of proof is transferred to the proponents of technological innovation (RR soybean).
Law	The preventive measures taken follow what the law prescribes.	The preventive measures are not in accordance with the law.
Science	Science as an instrument for objective decision-making. The "politicization of science" should be avoided.	Advocacy for a more rigorous approach to risk assessment through the application of EIA. Scientific uncertainty poses scientific and normative questions.
Participation	Decisions should be left to experts and scientists. CTNBio should have control of the risk management, communication, and analysis process.	Scientists and experts should communicate scientific uncertainty to other decision-makers. Decisions need to be transparent and open and involve civil society participation. Management and RA should be seen as separate.
Sustainable agriculture (SA)	Biotechnology enables more sustainable agricultural practices.	Tg soybean could pose a risk to the sustainability of ecosystems if stringent precautionary measures are not applied.
Risk of GM soybean	No specific risks other than the risks associated with conventional soybean. The greatest risks are associated with the commercial ban on soybean and its consequences for agricultural, economic, and scientific modernization.	International cases point to the risks associated with RR soybean. A commercial release of soybean represents new risks for the country in international trade, for the consumer, and for natural ecosystems.
Hazards associated with the risks	Hazards are associated with the trade ban. Loss of market of Brazilian agriculture and the undermining of scientific technical development.	Hazards are associated with the commercial release of RR soybean. Loss of market and greater risks in the consumption of conventional soybean (contamination). Greater environmental impact of RR soybean with the greater use of herbicides in agriculture and the existence of risks pointed out by scientific papers.
Benefits associated with risks	Commercial release brings greater benefits. Lower environmental impact with less water use and lower carbon oxide emissions and higher agricultural productivity, enabling greater gains for producers and lower costs for consumers.	The trade ban brings greater benefits. Safety for the consumer and natural ecosystems. Market acceptance for nonconventional soybean allows greater gains for farmers and for Brazilian agriculture more generally.

Source: Author

5.1 A World Where Everything Becomes Ideology

In the process of the commercial release of RR soybean, advocates of RR soybean inscribed their critics within the framework of a struggle that would be more ideological than scientific. The text *In Defence of Science*, by Bresser-Pereira (MS&T), presented at UNESCO in 1999, when the conflict began, provides us with some clues as to how the actors in favor of the commercial release of GMOs came to perceive their opponents' point of view. In a speech given at UNESCO in the period, Bresser-Pereira calls attention to the fact that the world is immersed in a "confusion between ethics and science, between ethical problems involved in science, and problems that are in all their aspects scientific" (Bresser-Pereira, 1999b, s/p). Therefore, it would no longer be religion and tradition that are being challenged today, but "reason itself and its result noblest – science." Bresser-Pereira gives us precisely the case involving the commercial release of RR soybean to illustrate a situation in which "reason" would be contested:

> A good example of what I am saying is what is happening in relation to genetically modified foods. There is almost a consensus in the scientific community that the individual GM food, which has been carefully examined by a national scientific committee, will not do any harm to health or the environment. However, what we see in many countries in Europe are NGOs and public opinion adopting a negative attitude to genetically modified foods […]. (Bresser-Pereira, 1999b, s/p)

Bresser-Pereira is referring here to RR soybean. And he takes this case to illustrate how, in various policy fields, especially those marked by resistance to GMOs, the world is becoming a place where "everything is transformed into ideology" (Bresser-Pereira, 1999b). It should be noted that this view was not restricted to the minister's vision. It also found support in the view of scientific organizations and scientists who throughout the conflict supported CTNBio's decision to release RR soybean for commercialization. Some years later, the president of SBPC, in a letter published in a widely circulated newspaper in the country, affirmed that part of "the culture against transgenics has obscurantist roots". An obscurantism that would not only damage science and the economy but also the "environment through the indiscriminate use of pesticides" (Raupp, 2008). This shows that opposition to the commercial release of RR soybean was not only identified as opposition to science but also to environmental sustainability in agriculture itself.

The thesis of obscurantist resistance by environmentalists was also expressed through a more subtle version, seeking support in a pedagogical discourse. To the extent that such resistance was not based on scientific knowledge, part of the solution to public criticism of GMOs would lie in the need to provide the public with more information about this innovation. Thus, the distrust for these products was not, as one scientist in favor of the release said, "based on facts, but on the individual's mistaken perception of the possible risk". And, although the issue brings us to the conflicting measures of the government, the main problems were associated with "the unpreparedness of the media to well inform their readers on cutting-edge science issues," as well as "the lack of scientific education of the average Brazilian population," being the distrust of GMOs due to the "limited knowledge of the subject and the little clarifying information conveyed by the media" (Lajolo, 2004, s/p).

But classifying GMO resistance as the result of an obscurantism alien to reason, or of a pathological state of misinformation among critics, tends to present problems when one realizes that a significant body of scientists opposed the release of RR soybean in the period. A critical position defended even by members of the CTNBio. This vision also found support in leaders such as José Luzenberg, an internationally renowned environmental leader in the RR soybean liberation period, Luzenberg presented a different diagnosis from that offered by Minister Bresser-Pereira. Lutzenberger (1999) pointed out that the "problem of transgenics is not technical; it is political!" This suggested that the conflict was being produced by the government itself. This makes it understandable that, during the period, congressman Fernando Gabeira interpreted the conflict not as a struggle of "ideology" (GMO critics) against "science" (CTNBio), but between "science against science." In this way, it is possible to see that proponents of RR soybean release generally saw themselves as the representatives of reason, associating their critics with a kind of obscurantism driven by some kind of ideological blindness. This stance in turn has not prevented the very government and scientists (from CTNBio) who defended the commercial release of RR soy from being accused, themselves, of obscuring their ideology by means of a technical-scientific discourse.

5.2 When Security Is Not Secure Enough

In the controversy over the commercial release of RR soybean, what differentiated critics and advocates of RR soybean was not just a simple acceptance of PP on one side and its rejection on the other. In controversies over GMOs, the polarization does not necessarily occur between those who acknowledge the existence of risks as opposed to those who claim that these risks are totally nonexistent, as it is difficult to find someone who does not recognize that GMOs can involve minimal risk to health and the environment (Lacey, 2006). The difference lies in the way these risks have been interpreted and in the way each side has come to interpret the necessary preventive measures.

In the literature on PP, there is debate as to whether it is a scientific principle or something distinct. But, as McGarvin (2001) notes, PP as a principle can be seen as a social norm to guide policy. Science, he continues, is simply an instrument for testing assertions about facts, not values. Similarly, von Schomberg (2006, p. 19) reports that the "its application involves deliberation on a range of normative dimensions which need to be taken into account while making the principle operational in the public policy context." The judgment that GMOs are beneficial or dangerous to the economy, to the consumer, and to health does not undo the normative nature of the judgments being made in each case. Thus, in the controversy over the release of RR soybean, actors have come to present different views of how they interpret the PP. In the case of the liberation alliance, there has been a tendency to interpret this principle in narrower terms. Thus, in this alliance, the PP was generally reduced to the realization of an RA, which, in turn, was perceived as being the responsibility of a select group of specialists linked to CTNBio. The message that was conveyed in the conflict was that the safety of GMOs should be dealt with only by biotechnology scientists. In this view, precaution tends to be reduced to a merely scientific issue, since preventive measures have been interpreted in this way. The discourse on commercial release of RR soyben, while presenting preventive elements, tends to present a suspicion of PP's most demanding views as an obstacle to scientific and economic progress when understood beyond these terms. That is, beyond the application of a RA.

In the case of the precaution alliance, headed by Greenpeace and IDEC, the appeal to PP presented itself in a different way. On its website, Greenpeace offered in the period a view on how it understands PP. For the organization, the principle was seen as a:

> [...] general rule in situations where serious and irreversible threats to health and the environment exist and require action to prevent such threats, even if there is not yet definitive proof of damage. This principle does not allow the absence of scientific certainty to be used to delay preventive action. (Greenpeace, 2006, p. 21)

At the expense of this Greenpeace view, and of other actors who have endorsed it, RR soybean liberation advocates have presented a minimal view of precaution. Greenpeace itself recognizes this when it notes that PP is usually translated as "a conventional risk assessment" (Greenpeace, 2006, p. 5). What, according to them, occurred in the process of releasing RR soybean in the country. The decisions of CTNBio, Monsanto, and parts of the government, throughout the conflict, reflected this very point of view and part of Greenpeace's criticism of the government is based on a critique of this position, which was perceived as reductionist. This is because the release of RR soybean was based on the performance of a conventional RA, considering this procedure as sufficient to legitimize such a decision. Such legitimacy would have a fundamentally legal basis. Hence, the defenders of the RR soybean release claimed that, if the environmental legislation required preventive measures and care in the release of RR soybean, these would be duly applied according to what the Brazilian environmental legislation informs. This perception was presented by Minister Bresser-Pereira (MS&T) himself in the period:

> [...] the National Congress approved the Biosafety Law and this law established the National Biosafety Technical Commission (CTNBio). [...] It is then up to the CTNBio to verify, on a case-by-case basis, whether or not a given product is likely to be approved for health and the environment from a biosafety standpoint. [The most recent example of the Commission's competence in implementing the Biosafety Law was the approval and regulation of the commercial use of "round up ready" transgenic soybean. This product was analyzed at length and finally approved by CTNBio. Therefore, the policy of the National Congress regarding transgenic products is being rigorously complied with. (Bresser-Pereira, 1999, p. 02)

The precaution alliance, in turn, was characterized precisely by considering the RA an insufficient instrument for decision-making on the commercial release of RR soybean. At the same time, it developed a critique of this supposed legal legitimacy that underpinned the government's decision to release RR soybean. In this case, the controversy over the soybean release involved two distinct ways of interpreting the risks present in the decision and the precautionary measures needed to address them. This is because the preventive measures associated with PP are often the subject of conflict. These actions can often range from determining measures for the restricted use of GMOs and the creation of requirements for the monitoring of plantings, as CTNBio did, to demanding the creation of more elaborate studies and requesting labeling, as civil society actors did, in particular, Greenpeace and IDEC.

5.3 Universalism and Equivalence in Risk Analysis

The controversy over the release of RR soybean is inextricably linked to the perception that different actors have established of the possible risks associated with its commercial release. The perception of risks in the use of RR soybean can be seen from two phases that complement each other in the dispute. At the beginning of the dispute, considerations about the risks associated with RR soybean proceeded in a projective and deductive manner to then incorporate a more empirical basis.[1] To the extent that the conflict over RR soybean was presented from a technological innovation that had not yet been implemented in the Brazilian agricultural system, considerations on risks tended to be projective, as they used future prognoses rather than statements on facts associated with the current use of RR soybean in the Brazilian context. At the same time, they were also deductive, as they draw conclusions about these impacts of RR soybean from other studies foreign to the Brazilian reality.

Both discursive alliances proceeded in this manner. CTNBio deduced that, since the RA conducted in the United States attested to the safety of RR soybean, the conclusions of these studies could be extrapolated to the Brazilian case. Monsanto's RA, validated by CTNBio, came to express a universal content, where conclusions on the impacts of RR soybean were then deduced independently of the social, economic, and ecological context in which it would be deployed. In doing so, CTNBio offered an unusual interpretation of the case-by-case principle, which indicates that GMOs must be assessed from the particularities of the context in which they are placed.

The purpose of "case-by-case" studies is to treat every release as "unique" (Myhr & Traavik, 2002, p. 79). CTNBio, in turn, treated RR soybean releases as equal or equivalent to those in the United States and used an RA that did not examine the

---

[1] This empirical phase of the disputes about the risks associated with RR soybean concerns the arguments on the impacts of RR soybean on Brazilian agriculture. After the release, both IDEC and Greenpeace started reporting research data that would indicate that the use of RR soybean would be causing an increase in the use of pesticides in the country. See, for example, IDEC (2008) and Greenpeace (2008). On the other hand, actors aligned with the liberation discourse began to indicate the opposite: the decrease in the use of pesticides, using foreign cases (e.g., China) as a basis in their arguments (Transgenics, 2001).

environmental peculiarities of RR soybean releases in the Brazilian case. This was due to the application of the Principle of substantial equivalence (PSE), which will be addressed below. Rather than the context and systemic relationships that a GMO may establish with its environment, the RA guided by this principle starts to examine the safety of the organism from its intrinsic chemical properties.[2]

Another point to consider in the perception of RR soybean's risks in the conflict is what Lacey (2006) calls the modern valorization of control. This valorization refers to "a set of specifically modern values connected with the control of natural objects, which relates to the expanding reach of technological control, without its value being systematically subordinated to other ethical and social values (...)" (Lacey, 2006, p. 19). These values are often fundamental to the very way in which GMO-related risks are perceived. The greater the confidence in control, the greater the perception of safety. Where this set of values is present, it will induce a perception of GMOs as informed in an exemplary manner by scientific knowledge (Lacey, 2006, p. 35). GMOs, as products of a *Sound Science*, come to be regarded as the result of highly reliable and respectable knowledge, putting aside the merit of criticisms that raise doubts about the reliability of that same knowledge. Thus, they transfer the burden of proof to the critics of GMOs.[3]

An enunciation of this kind of valorization can be seen in the words of Luiz A. B. de Castro, former president of CTNBio, who in the period of the release of RR soybean declared that:

> [Genetic engineering is just one more method of genetic improvement, similar to many others that have been used in the past, such as gamma radiation, chemical mutagens etc., and which were not subject to labelling, even though classical breeders had less control over the genes being manipulated. With the advent of genetic engineering we have absolute control over the manipulation of these genes. (de Castro, 1998, s/p)

The statement offered by Bresser-Pereira, claiming that the "product was long analyzed and finally approved," follows the same line in linking the legitimacy of the

---

[2] In Chap. 6, this question will be taken up again which addresses the issue of labeling in the process of release of RR soybean. It will become more evident in this part of the work how the safety measures used to define the commercialization of RR soybean have reproduced the precepts of the American regulatory process for GMOs.

[3] Sound Science is not synonymous with "good science" and the term does not have a specific definition in the scientific field. The term refers to a type of argument that is used in political controversies where one seeks to invalidate preventive measures by offering a skewed reading of scientific knowledge. In his book *The Republican War on Science*, Chris Mooney notes that for those who position themselves to the right of the political debate, Sound Science means demanding a "higher burden of proof before action can be taken to protect public health and the environment." As Mooney (2006) indicates, the expression does not necessarily suggest a scientific position reliable, because those who take this point of view do not always have an adequate reading of the "state-of-the-art" knowledge. As an argumentative artifice, the appeal to Sound Science is usually applied by those who oppose, as he notes, the precautionary principle, opposition that tends to operate, more in ideological than scientific terms. He says: "If Sound Science stands in contrast to the 'precautionary principle', then that contrast, by definition, must occur within the sphere of policy, not science" (Mooney, 2006).

5.3 Universalism and Equivalence in Risk Analysis

decision to the reliability of the knowledge (RA) employed to ground the commercial release. Since the decision was based on a Sound Science view of science, there would be no reason to entertain any doubts in the RR soybean release process. It should be noted that these values are also associated with the belief that intensive conventional agriculture is the only viable alternative to modern agriculture. An aspect that, in the case of RR soybean, will fuel a conflict between RR soybean proponents and sectors that have advocated alternative agricultural methods, as examined in the first chapter. This modern valorization of control did not emerge in the considerations that were made about environmental risks, because these became rightly underestimated. Rather, these values began to emerge in the debate on the safety of RR soybean, because it is in this dimension of the conflict that the assumptions of control are expressed. Assumptions which, by the way, have come to be questioned by the precautionary discourse. Among these assumptions is the claim of the existence of a "consensus" and "scientific proof" on the safety of RR soybean, the validity and reliability of the Principle of substantial equivalence (PSE), the possible success of soybean liberation in other countries, and other issues associated with these points. All of these claims have been used to legitimize the commercial release of RR soybean and all of them the precaution alliance has sought to contest by relying in some way on its own scientific authority, thus using its own science to counter the science of its opponents.

The Principle of substantial equivalence (PSE) has had a major influence on how the release discourse alliance has come to interpret the risks associated with RR soybean. In the RA submitted by Monsanto, the company reports that "the results of the technical and scientific studies conducted with RR soybean in Brazil in other countries demonstrate its environmental and food safety" and that part of these conclusions can be drawn from the "substantial equivalence of Roundup Ready soybean to non-modified soybean" (Monsanto, 1998 apud Cezar, 2003, p. 79). At the same time, as Cesarino (2006, p. 81) points out, CTNBio "accepted Monsanto's arguments that the studies it presented demonstrated the 'substantial equivalence' of Roundup Ready Soybean relative to genetically unmodified soybean," which meant that the risks associated with RR soybean then came to be considered as "the same as those associated with conventional soybean."

But it is possible to say that the interpretation of the PP had occurred with some caveats by CTNBio, since if the risks involved in the use of RR soybean were exactly the same as those linked to conventional soybean, the preventive measures announced by CTNBio would lack implementation justification. The very preventive measures taken by the commission, such as the request for the creation of a biosafety monitoring committee, seem to indicate that even CTNBio itself recognized that there was no full equivalence between RR soybean and conventional soybean. And that apparent contradiction was noticed by judge Antonio Souza Prudente, who, in the public civil suit filed by IDEC against Monsanto and the Federal Government, points out that "if the product is really safe, there is no reason to submit it to monitoring, with rules that reveal the danger of environmental damage" (Justiça Federal, 2008). In other words, CTNBio would have used a conventional RA to examine the risks of RR soybean, but ended up making prescriptions that denote the unique character of the product.

It is for this reason that Lacey (2006, p. 214) calls attention to the fact that the PSE has no relevance to the discussion on environmental risks. If some products are considered as "equivalent" to others already on the market, environmental impact assessments (EIAs) become dispensable tools for risk assessment, since this principle induces us to think about the consequences of GMOs from the experience of previously known substances. The risks associated with RR soybean were considered, then, as already partly known, since they are deduced from experience with the use and marketing of conventional soybean, which has turned RR soybean and conventional soybean into equivalent products. And so have the assessments of their risks.

Rather than being associated with the use of RR soybean in agriculture, the greatest risks lay, for RR soybean liberation advocates, in the possible legal ban on the commercial use of the product. Fears were directed at the undermining of the country's agricultural modernization and the indirect implications that the measure might have for the development of science. In the liberation discourse, economic losses were integrated with scientific losses due to the common projects established among agbiotech companies, universities, and government. RR soybean was not only seen as similar to traditional soybean, but was considered to bring lower environmental risks and new economic opportunities. Unlike traditional soybean, RR soybean was therefore seen as even closer to sustainable agriculture. The following passage from the president of the SBPC illustrates this perception:

> In terms of environmental gains, it is estimated that in ten years the adoption of transgenic soybean plantations will save 42.7 billion liters of water and 300 million liters of diesel oil, with a reduction in emissions into the atmosphere of approximately 1 billion tons of carbon dioxide, equivalent to planting 6.8 million trees. For cotton and soybean, a reduction in the use of more than 77,000 tons of pesticides is estimated. Our industry is the most affected by the entry of pirated seeds, due to the delay in adopting transgenic seeds. (Raupp, 2008, s/p)

This view reinforces Bresser-Pereira's thesis on the trade-off examined in the previous part. It suggests a "positive sum game" where all political and economic actors would gain benefits from the economic release of RR soybean. It puts in other terms the former minister's statement that "there are no losses in terms of biological security and environmental protection, as a counterpart to economic development gains" (Bresser-Pereira, 1999, p.1).

5.4 Epistemological Uncertainty and the Precautionary Discourse

What Myhr (2007) calls epistemological and methodological uncertainties help to understand the controversy raised by the precautionary discourse. Epistemological uncertainties are associated with the imperfection of current scientific knowledge

5.4 Epistemological Uncertainty and the Precautionary Discourse

about the benefits and potential adverse effects of GMOs. This uncertainty tends to arise because of the novelty of an innovation, as in the case of RR soybean, or the inherent variability or complexity of a system under consideration. Methodological uncertainties are related, in turn, to the choice of methods for assessing the risks associated with the use of new products brought by the innovation.

The precaution alliance will seek to challenge the discourse of RR soybean release by acknowledging these two sets of uncertainties. But just as CTNBio relied on a deductive judgment from assessments drawn from the RA conducted in the United States, the precaution alliance relied on the perception of risks from international cases to also sustain its opposition in the conflict. Among the epistemological uncertainties indicated by these actors in the conflict was (a) the indication of the alteration of the chemical composition of conventional soybeans that could occur if they came into contact with RR soybean; (b) the acknowledgment made by Monsanto itself, which would have, in 2000, reported in Europe the existence of two unforeseen gene fragments in Roundup Ready soybeans; (c) the case of scientist Arpad Pusztai who, in a 1999 interview, raised questions about the methodology used by Monsanto to evaluate the risks associated with RR soybeans[4]; and (d) the conflicts involving the regulatory process in other countries that have used the Principle of substantial equivalence (PSE).

As IDEC and Greenpeace have pointed out, both the US Food and Drug Administration (FDA) and the US Environmental Protection Agency (EPA) have faced lawsuits challenging the authorizations in the country. In addition to accusing the regulatory agencies of not complying with technical formalities, critics in the United States also pointed to the nonexistence of the claimed "technical consensus" in the clearance of RR soybean (IDEC, 2003a, b).[5]

But this lack of technical consensus was apparently not a particularity of the US regulatory process. The CTNBio itself, since the release of RR soybean, has been permeated by internal conflicts that point to members' opposition to the commission's regulatory policy being implemented at the time. In 1999, biologist and ecology PhD Maria Alice Garcia, a CTNBio member, left the commission because she disagreed with the commission's current policy. During this period, she herself confided that the "opinions reached a degree of divergence" within the CTNBio, which caused her deep emotional distress (Garcia, 2008 apud Divergence, 1999, s/p). In 2005, Lia Giraldo, a sanitarian physician and environmental specialist, until then a member of the CTNBio, also resigned from the commission for similar reasons. In a letter addressed to the MS&T, of the Environment, and to the CTNBio itself, the researcher criticized the commission's procedures. In the letter, she criticizes the "belief in a science of monocausality," which would be impregnating the commission's decisions. In opposition to this type of vision, she calls attention to the fact

---

[4] This scientist criticized the methodology used by Monsanto and, within it, the use of animal adults used in the tests performed by the company and the lack of measurement of the organs affected by the presentation of genetic material.

[5] In October 2003, IDEC filed a public civil action against the Federal Union and the state of Rio Grande do Sul, in which it exposes these arguments. See IDEC (2008).

that "we are dealing with complex issues, with many uncertainties and with consequences over which we have no control, especially when it comes to the release of GMOs into the environment" (Augusto, 2005, s/p). This report demonstrates that the technical consensus was nonexistent even within the CTNBio and that the commission was invaded by internal conflicts in its evaluations on GMOs. These conflicts were only reduced with the departure of those members who were critical of the internal procedures adopted by the commission.

Added to those arguments is the criticism to the approval of RR soybean in Brazil by studies conducted by the company itself which, according to IDEC and Greenpeace, were "obviously all favourable to its intention, without being scientifically validated" (IDEC, 2008, s/p). As these organizations noted, the Brazilian case presented an even more serious situation, since the studies presented by Monsanto did not even make reference to the specific environmental conditions of the country. IDEC supported this argument based on the evaluation made by the former president of CTNBio, Dr. Glaci Zancan, who indicated that "it is essential to repeat the tests with seeds cultivated in the country" and that the data should be confirmed "by teams independent of those hired by the company as a way to ensure the reliability of the results" (Zancan apud IDEC, 2008, s/p), which, despite this scientific advice, did not occur in the decision to release RR soybean.

With regard to the risks and dangers posed by GMOs, many academic papers point to the lack of a definitive consensus among scientists on the issue. Minister Bresser-Pereira himself partially acknowledged this fact when he said that, regarding the safety of RR soybean, "there is almost a consensus in the scientific community" (Bresser-Pereira, 1999, p. 2). The accusation that resistance to RR soybean reflects an obscurantism of critics, made by the minister himself at the beginning of the conflict, then, conceals two important points. It conceals the fact that considerations of the safety and risks posed by GMOs have been and still remain today a focus of dispute in the scientific field. In doing so, it also hides the fact that resistance to RR soybean, by organizations such as Greenpeace and IDEC, was based on studies and scientific information that contradicted the claims made by the release discourse alliance.

5.5 CTNBio, Regulation and Scientific Uncertainty

The European Court of Justice, in its interpretation of PP, reports that preventive measures cannot be based on purely hypothetical considerations or conjectures that have not been scientifically substantiated. There must be a basis that gives a scientific plausibility for the application of the principle. Some kind of basic scientific knowledge, which offers a threshold of scientific plausibility, is required in its application (Sadeleer, 2007, p. 20). To leave the hypothetical framework, the recognition of the existence of scientific uncertainty as a condition for the application of the PP becomes necessary. The application of the principle in the decision-making process requires, then, that scientific uncertainty becomes explicit (Myhr & Traavik, 2002,

5.5 CTNBio, Regulation and Scientific Uncertainty

p. 74). Scientists and risk analysts are responsible for recognizing and communicating this scientific uncertainty to the decision-making sphere. To the extent that it is not recognized, or discarded, there is the risk that the application of PP is annulled by the absence of this type of communication. Hence, PP, and its relation with scientific uncertainty, ends up conferring to scientists and specialists an important role in risk communication in the decision-making process.

In the Brazilian case, the fact that CTNBio conducted an RA can be perceived as the expression of recognition of the existence of a minimum level of scientific uncertainty that started to be considered in the decision-making process. The initial uncertainty framework was faced by the application of the RA that was presented by Monsanto. This basic scientific knowledge, represented here by the RA, seems to indicate that CTNBio itself recognized the existence of uncertainties involving the commercial release of RR soybean. But if the commission initially recognized a minimal level of uncertainty, it subsequently grounded its discourse in a series of assumptions that served to precisely disregard it, thus preventing it from being considered within the framework of certain more robust interpretations of the PP in which scientific uncertainties associated with GMOs are seen as being of a new type. In confronting the precaution alliance, the liberation discourse has come to rely on the scientificity, reliability, and sufficiency of the RA to justify its decision to release RR soybean for commercial use. It postulated that the RA represented the "best available scientific knowledge" and then claimed that scientific uncertainties regarding RR soybean were no different from uncertainties in the use of conventional soybean. Some of its stakeholders also postulated the existence of a "scientific certainty" and "scientific consensus" that would validate perceptions about the safety of RR soybean. The experience in other countries, somehow, would give even more validity to those claims. Bresser-Pereira's letter, written to the MS&T in 1999, makes several of these points clear. In his letter, Bresser-Pereira does not fail to recognize the existence of scientific uncertainty, but tends to reduce it to a type of uncertainty that would be equivalent to the uncertainty associated with the usual innovations of the economic process. Everything indicates that CTNBio had the full backing of the president of the MS&T for the release of RR soybean from this perception of the state of the art of knowledge on the risks of RR soybean.

While pro-liberation actors saw the RA as a dissipator of scientific uncertainty, the precaution alliance saw it as a way to hide it. Unlike those who defended the commercial release of RR soybean, the precautionary discourse went on to affirm and communicate the existence of this uncertainty to public opinion. It did so, first, by indicating the epistemological uncertainties related to the risks associated with the use of RR soybean, as previously examined. But it then sought to point out the inadequacy of the methods used to assess these risks. Since risk perception cannot be dissociated from the knowledge that is applied to define them, the debate over RR soybean then turned to the validity of the knowledge to define risk and safety in this case.

Supported by scientific studies, IDEC then pointed out that there were not enough experimental studies on the potential adverse effects of GM foods to justify their release (IDEC, 2003a, b, s/p). Among other arguments presented by the organization to contest the use of RA, and point out the methodological uncertainties, are the statements that (a) the Principle of substantial equivalence (PSE) is a pseudoscientific concept, since it advocates a limited number of tests for the determination of the composition and toxicological potential of products, (b) the toxicological evaluation also excludes a series of toxic substances that are not selected in the tests, (c) the analyses inspired by the Principle of substantial equivalence (PSE) would then not allow the evaluation of a series of toxicological differences between different types of soybeans, and (d) the release of RR soybeans in other countries occurred in studies presented by Monsanto itself, serving the company's own interests. The RAs should not, IDEC argued, be carried out by the requesting companies, but by independent research groups. The Committee for the Defense of Consumers, the Environment, and Minorities of the House of Representatives was the first to draw attention to some of these points back in 1999. In a report produced by that commission in 1999, it called attention to the fact that "The dossier presented by the proponent contains only, and only, information about the soybean in question when cultivated in the United States," emphasizing then that it considered "this level of information insufficient for taking a decision for what might be called deregulation of this product in Brazil" (Brasil, 2003 apud Pessanha & Wilkinson, 2005, s/p). All these arguments were supported by scientific papers and the opinion of Brazilian and foreign scientists.[6] It is no coincidence that the precaution alliance found support in different segments of the scientific field and that these, in some way, offered scientific criticisms similar to those made by civil society organizations that acted in the conflict more directly.

5.6 Does the Absence of Evidence Indicate the Absence of Risk?

The debate about the limits of RA brings us back to the question about "What?" must be proved and "Who?" is addressed to such responsibility. "This is because PP debates are generally structured on the basis of a conflict in determining the burden of proof in proving risks. In this conflict, proponents of PP tend to see GMOs as "guilty until proven innocent," while proponents of innovation tend to see them as "innocent until proven guilty" (van den Belt & Gremmen, 2002). In the RR soybean controversy, this polarization has played out with each discursive alliance holding its opponent responsible for proving the dangerousness or safety of RR soybean. In this debate, modern values of scientific–technical control tend to influence the debate in some way. These present an important element on disputes involving RR soybean because where they are present, they will induce a perception of GMOs as informed in an exemplary manner by scientific knowledge (Lacey, 2006, p. 35). In

---

[6] The considerations about Principle of substantial equivalence (PSE) were based on the work of Brunner et al. (1999).

5.6 Does the Absence of Evidence Indicate the Absence of Risk?

doing so, the trust placed in technological innovations (GMOs) enables proponents of the innovation to feel empowered to sustain its present and future use, removing from critics any scientific credentials to oppose the innovations. These values and the trust in biotechnological science are what allow one to defend the thesis of the safety of GMOs and then subsequently shift the burden of proof to the critics of these products.

This is why the defenders of the commercial release of RR soybean saw as normal the demand they addressed to the critics in the conflict for them to present the scientific evidence that could prove the dangerousness of RR soybean. This discursive structure can be seen, for example, in the letter written by the president of CTNBio in the period when he states that the "The campaign [against GMOs] has not presented to the world, in more than a decade, even one scientific evidence that justifies the condemnation of transgenic they preach" (de Castro, 2008). This same argument was also put forward by a scientist in favor of the release of RR soybean, stating that "the safety of these products is scientifically proven and no relevant argument has been raised to counter this statement" (Pavan, 2008, s/p). These arguments have become quite common in the conflict to provide authority to the argument for the commercial release of RR soybean.

The different elements that support this assumption about the safety of OGM soybean must be considered. Obviously, if the Principle of substantial equivalence (PSE) allows concluding that risks are equivalent, then the safety standard associated with each type of soybean will also be equivalent. This premise is also supported by the thesis of Bresser-Pereira (1999, p. 2), who, as already seen, stated that "there is almost a consensus in the scientific community" that would confirm such safety. But there is an additional fundamental assumption that supports this point of view. It rests on the thesis that, given the absence of substantive data and information on the adverse effects of GM soybean, this absence would confirm the presence of GMO safety. Safety, in this case, is justified not by the existing substantive information on the risks under discussion, but by the lack of information and studies on them. As Walter Colli, CTNBio president, points out: "After so many years, widely planted and consumed in various countries, we have not heard of any ecological disaster or strong allergies in people and animals" (Colli, 2004, s/p). The absence of substantive information on the possible adverse effects of RR soybean would be to prove, therefore, its safety.[7]

This discourse points to an argument that can be found elsewhere as well. Myhr and Traavik (2002), examining the regulatory process associated with the commercial release of GMOs in other countries, note that:

> It has been argued that there is not enough evidence to reject the hypothesis that GMO and GM food is safe. The fact is, however, that experiments designed to clarify potential adverse effects on health or the environment are nearly absent in peer-reviewed journals. Hence,

---

[7] This argument neglects several points that are commonly raised by critics of the GMOs. These include the recent use of GMOs in agriculture, the precariousness of the methodologies created to evaluate their effects, and the complexity and temporal dimension that may be involved in the long-term consequences associated with this type of technological innovation.

scientists and regulators are more interested in avoiding false positives (type I-error), than false negatives (type II-error). (Myhr & Traavik, 2002, p. 80)

When RAs are permeated by a number of uncertainties, risk assessors and regulators may make two types of mistakes. Given the uncertainties, they may wrongly classify a substance as hazardous (false positive) or they may, on the other hand, make the opposite error: failing to identify the hazardousness of a substance by classifying it as safe (false negative). Thus, such an option leads regulators to conclude that to the extent that research does not indicate any significant adverse effects of GMOs, they can be classified as safe (Myhr & Traavik, 2002). In other words, unable to support such decisions within a solid framework of scientific certainty, regulators prefer to make one type of error (false negative) rather than another (false positive).

CTNBio's decision to release RR soybean exemplifies precisely this pattern of understanding. To the extent that there is no substantive evidence to inform the harmfulness of RR soybean, this would seem to indicate that CTNBio sought to avoid a false positive. To the extent that the harmfulness of RR soybean can only be assessed from its direct use or application, given the imprecise informational picture, banning the use of RR soybean would not make it possible to examine the adverse effects of its application, as such a decision would provide no informational basis for drawing any kind of conclusion. Banning its use, in turn, would imply identifying it as dangerous. But this attempt to avoid a false positive, as Cranor (1993) points out, may be consistent with scientific rationality, but not necessarily with the purposes of the regulation process. If in the scientific field the attempt to avoid false positives tends to be central, according to Cranor, in regulation false negatives are of greater importance. In the regulation of potentially dangerous substances, the health of the population is the main objective of regulation, unlike science, which may privilege other values. Thus, "what is typically of lesser concern in purely scientific inquiry is of much greater concern in regulatory inquiries" (Cranor, 1993, p. 26). Unlike CTNBio, the precaution alliance chose to legally activate PP because it presumably implies a differentiated regulatory stance, since the application of PP suggests, as Myhr and Traavik (2002) point out, the prevention of false negatives.

Perhaps this is why Lia Giraldo, a former member of CTNBio, pointed out in 2005, in a letter requesting her resignation from the commission, that CTNBio "is not a body that fosters research or graduate studies or an editorial board of an academic journal" (Augusto, 2006). In this passage, the former CTNBio member suggests that the considerations on safety of GMOs in the regulatory process could not be the same as in the scientific field, precisely to indicate that the prevalence of scientific rationality should not override regulatory rationality. The latter presumes greater care that the former may eventually disregard.

This differentiated perception of the precaution alliance was based on a questioning of this assumption of the absence of information as an indication of the safety of RR soybean. Newton de Lucca, a judge at the Federal Regional Court for the Third Region, aligned with the precaution alliance discourse, for example, held that the

5.7 Beyond Risk Analysis?

possible absence of scientific proof of harmful effects from the use of GMOs in agriculture did not necessarily authorize marketing the product as if it were considered harmless to human health:

> It is exactly the opposite that must be observed, that is, scientific certainty must exist in the sense that there is no possibility of damage to the environment, in accordance with the so-called precautionary principle now required as a rule of international law. (Lucca, 20--? apud Porto, 2002, s/p)

From this viewpoint, the absence of data and information on the hazardousness of RR soybean would not be indicative of its safety. This does not necessarily mean that it is dangerous for the environment and human health. But such information frame should be taken rather as an indication of doubts and uncertainties still existing from that perspective. Knowledge should occur, then, on the basis of building on some kind of existing basic scientific knowledge and not by the lack of it. This lack of knowledge was seen to emerge in the contestation that was made to the RA, which should precisely exercise this role of reducing or communicating uncertainty. However, once the RA was challenged, knowledge about the adverse effects of RR soybean was seen as nonexistent or unreliable. This framework shows that scientific uncertainty was seen as a basis both for sustaining the commercial release of RR soybean and also for challenging it.[8]

In the conflict, other criticisms were also important for the precautionary alliance to shift the burden of proof to proponents of RR soybean release. The lack of studies conducted ended up serving to invalidate the universalism associated with the RA presented by Monsanto. Since the test regarding environmental risks had not been conducted in the context of Brazilian ecosystems, this allowed the precautionary alliance to assume that the priority burden had not been assumed by producers and users of these innovations (Lacey, 2006). The questioning of the Principle of substantial equivalence (PSE) served the same purpose. This principle, as already seen, tends to redirect the burden of proof to critics of GMOs, who, based on it, would presumably have to demonstrate the nonexistence of equivalence. But questioning the premises of this principle tends to redirect the burden of proof back to the proponents of innovation. Proving nonequivalence would imply presuming, for example, that it can be established, or that it is in any case valid for assessing the risks associated with RR soybean. This is why the challenge to the precaution alliance did not occur by trying to establish nonequivalence, which might not be possible for the case at hand.

---

[8] As Cesarino (2006, p. 81) points out for the Brazilian case: "the lack of a scientific definition unequivocal about the risk is argued for both the release of the technology and its moratorium: the absence of definitive proof of its harmfulness authorizes its release, just as the absence of definitive proof of its harmlessness authorizes nonrelease (or, at least, the requirement for further studies)." As "zero risk does not exist, there will always be a margin for both readings, a since science is not able to definitively establish either the harmfulness of the product or the safety of the product" (CESARINO, 2006, p. 81).

5.7 Beyond Risk Analysis?

The RA has been seen as providing a means of acquiring basic knowledge for PP application. At the same time, it has been considered as a means of accessing the uncertainties existing in the decision-making process for the release of GMOs (Sadeleer, 2007, p. 21). The RA, in this case, has been considered the "scientific method of confronting and expressing uncertainty in predicting the future" (Lohani et al., 1997, p. 55), being that the "Uncertainty about the exact nature, frequency and magnitude of those consequences is inherent to both EIA and risk assessment" (Lexer et al., 2008, p.12). What makes the EIA and the RA are founded on very similar concepts and present the same objectives, allowing that the RA, in certain cases, is often used in EIA or even perceived as a study equivalent to it. The similarity that can be established between these methodologies may also depend on cultural factors. Unlike what occurs in the United States, the application of RA, as a type of EIA, tends to be rare at the European level (Lexer et al., 2008).

Brazilian legislation and the legal sectors that align themselves with the precautionary discourse alliance have followed this last type of interpretation. However, the actors belonging to the discursive alliance of liberation began to see the RA as a methodology equivalent to EIA, something that occurs in the United States. This is made clear by Simone H. C. Scholze, a CTNBio member at the time of RR soybean release:

> The environmental risk assessment and control process from the point of view of GMO biosafety, which must be strictly followed by the applicant for authorization to carry out experiments with transgenic organisms, is minutely defined in CTNBio's normative instructions no.3/96 and no.10/98. These regulations contain detailed rules for the evaluation and control of environmental risk, as well as the risks to human and animal health arising from the use of transgenic organisms, whose content and criteria are substantially equivalent to an environmental impact study, although not called such. (Scholze, 1999, p. 34)

This same equation is what the former president of CTNBio did, when he states that "[critics of GMOs] propose environmental impact studies that, in practice, have already been done all over the world, including Brazil" (de Castro, 2008, s/p). In short, the demand by critics of RR soybean release for an EIA to validate the safety of RR soybean would be incoherent, since equivalent or similar studies have already been conducted. For the case at hand, the RA is taken to be equivalent to an EIA. Or it is perceived as being a risk assessment method that in terms of information and methodology could be seen as similar. The equivalence thesis was apparently used not only to examine the different types of soybean but also to establish a similarity between the different types of EIA methodologies in dispute. EIAs would not be necessary because they would already have been carried out, the RA being their equivalent.

However, despite the clash that has taken place on this topic, both in Europe and in the United States, RA has been the focus of increasing criticism. RA does not always provide accurate and reliable scientific knowledge about cause-and-effect relationships and does not, most of the time, eliminate the existing uncertainty and knowledge deficit. In many cases, it is claimed that this void tends to be filled with assumptions and intuitions of risk experts themselves. As Fischoff et al. (1981, p. 33) note: "many risk problems force experts to go beyond the limits of the available data and convert their incomplete knowledge into judgments usable by risk assessors." The evaluations themselves may be influenced by external pressures, following narrow interests or, depending on how demanding they are, they may overly bind the decision-making process, restricting decisions to specific issues. When too narrow, the scope of the RA can exclude the use of different RA methodologies and the different types of knowledge that can be used for this purpose. And to the extent that RA is seen as a strictly scientific procedure, it tends to hinder the participation of a diverse range of stakeholders.[9]

Many of the criticisms that the precaution alliance has addressed to the RA echo some of these points. The arbitrary assessments of scientific uncertainty, the pressures to release RR soybean, the exclusion of areas of knowledge important to the RA, and the narrow technocratism that guided CTNBio's risk assessment process were common criticisms made by the precaution alliance in the conflict. In 1998, when CTNBio issued its opinion releasing RR soybean, the Report of the Committee for the Defense of Consumers, the Environment, and Minorities of the House of Representatives pointed out that "there was undeniable haste in the decision taken by CTNBio" with clear signs of "unquestionable eagerness in the manifestations of the authorities involved, for the release" (Brasil, 2003 apud Pessanha & Wilkinson, 2005, p. 63). Of special importance here was the letter written by Ely Geraldo da Silva Augusto (2005), which also points to several of these problems existing within the CTNBio.

The use of the RA to release the use of GMOs has then raised a number of controversies. If such analyses are necessary, what types of studies need to be conducted? Is the RA sufficient and satisfactory to address the new risks associated with GMOs? What must be done when the law does not specify the appropriate methods to assess these risks? In the absence of consensus, how should the RA deal with diverging scientific opinions? The case of RR soybean shows that there was no consensual basis for many of these questions. Critics of the release insisted that RR soybean required an EIA and that such an EIA had not been conducted. The alliance favorable to the commercial liberation of the GMOs, on the other hand, has done the opposite, making the RA an equivalent of EIAs, deeming such a tool sufficient and legitimate for the decision.

Brazilian environmental legislation stipulates the need to conduct EIAs for the release of GMOs, but does not specify which EIAs need to be conducted. The Federal Constitution of October 5, 1988, establishes in Article 225 that the public authorities have the duty to defend the environment, being responsible for "requiring a prior environmental impact study for the installation of works or activities that may cause environmental degradation" (Brasil, 2003 apud Pessanha & Wilkinson,

---

[9] For an examination of these criticisms, see Sadeleer (2002). For an examination of the limitations of the RA produced by regulatory bodies in the United States, see the National Research Council (1994).

2005, p. 51). The Brazilian environmental legislation, thus, does not provide precise definitions of these points. It has defined the role of certain bodies and the need to conduct EIAs, but does not specify the content and characteristics of EIAs for the release of GMOs. It does not indicate, for example, whether the RA can be seen as studies equivalent to EIAs and how these methodologies can be used in the GMO regulatory process. The profiling and scope of these methodologies depends on the decisions of regulatory bodies. For this reason, the solution to these impasses has been to transfer control over the carrying out of the RAs to CTNBio itself, preventing other government agencies from contesting the choice of methodologies for assessing the social and environmental impacts of GMOs.

While this may mitigate conflicts, it may lead to a loss of diversity in the use of RA methodologies in assessing the risks associated with GMOs. To the extent that conducting EIA was seen by the precaution alliance as a way to overcome the problems posed by RA, a return to the problems to which this vision refers us is necessary. EIAs cannot be considered as more scientific, since both EIA and RA depart from the usual standard of science. As Beattie (1995, p. 110) indicates, EIAs are not produced to test and refine explanations, but are produced to predict potential impacts. RAs are no different in this respect. Moreover, EIAs are rarely conducted by independent researchers, but by experts working in the service of corporations and governments. While the criticism of RA by Monsanto focused on its possible bias, there is no guarantee that EIAs conducted by government agencies, for example, do not suffer from the same problem. To the extent that EIAs are directed at potential future impacts, it is common that these studies present serious gaps in terms of information and that they also work from simplistic assumptions due to the indeterminacies that surround the distance between the present and the future in the assessment of impacts.

EIAs are also steeped in value assumptions. Cost–benefit analysis, for example, a methodology that is commonly used in EIA, is based on a "set of value assumptions about that nature of human motivation and behavior " (Beattie, 1995, p. 111). EIAs are not only associated with normative judgments, but can also be considered as essentially political. To the extent that all modernization projects involve some sort of distribution of costs and benefits, "EIAs are part of a decision-making process that has distributional impacts" (Beattie, 1995, p. 112), which makes them, as this author reminds us, "political." Because of the way that RA and EIA are traditionally understood and because of the factors considered in the fourth chapter, the distributive issue tends to be neglected. Neither RA nor EIA, therefore, escapes questions involving values and politics, and neither should be seen as more scientific, if one wants to understand the possibility of transcending problems of this kind. These elements are integral components of these methodologies and as such cannot be eliminated, only acknowledged or hidden.

To conclude this analysis, an important point of the conflict that was neglected in the previous pages will be analyzed but which deserves a brief assessment at the end of this chapter. The RR soybean liberation process raised a series of conflicts involving civil society participation in the decision-making process. A conflict that did not extend only to the issue of institutional accountability involving other government

5.7 Beyond Risk Analysis?

agencies but also involved issues linked to civil society participation in the decision to release RR soybean. In the conflict, the precaution alliance criticized precisely what it perceived as a closure of the decision-making process that was reflected by CTNBio's excessive control over the process. In her letter, former CTNBio member Lia Geraldo da Silva Augusto points out that it was a "big mistake to remove from regulatory and inspection agencies the powers to analyze and decide" (Augusto, 2005, s/p) on release-related issues. She also makes the same judgment, considering even bizarre, the situations in which members of civil society were prevented from participating in decisions.

A clear sign of the control that the CTNBio began to have throughout the process can be seen in the controversies generated in the elaboration of Conama Resolution 305, of 12/06/2002, which "Provides for Environmental Licensing, Environmental Impact Studies and Environmental Impact Reports on Activities and Undertakings with Genetically Modified Organisms and their Derivatives". The resolution established in its article 4, paragraph 3, that a:

> [...] the risk assessment of GMOs is the responsibility of CTNBio and shall be considered by the competent environmental agency as part of the environmental RA process, which shall be complemented with the risk management and communication, considered additional requirements and procedures of legal and exclusive competence of the competent environmental agency. (Conama, 2002, p. 717)

Those who drafted the first draft of this resolution drew a distinction involving management, communication, and RA. The management process was conceived as involving these three main phases. CTNBio would be responsible for carrying out the RA, while policy decisions would be left to the new committee. However, article 4 of the aforementioned resolution ended up displeasing government representatives and, above all, members of the ministries of agriculture, science and technology, and especially CTNBio. In this process, CTNBio considered that this competence would be "exclusively theirs" (Marinho, 2003, p. 89). That is, the liberation alliance claimed exclusive control over the three management processes, thus producing a process of centralized decision-making.

CTNBio's control over the RR soybean release process represented a framework in which management and RA became concentrated in a single committee. In this framework, a policy prescription in terms of values (the decision to commercially release GMOs) was removed from what was considered to be a judgment on facts provided by the RA. But the attempt to separate the "management" from the "RA" seeks precisely to prevent the political decision from being subjected to a technical or scientific decision (RA). As noted earlier, the "facts" from EIAs and RAs refer us to normative judgments and distributive political issues that underlie these studies. To assume that these "facts" are "pure" implies not only distorting the type of information at hand in these cases but also blind acceptance of the normative judgments that are embedded in these studies without submitting them to any kind of deliberative mediation. Science and technique, in these cases, offer answers to questions that are not the appropriate means to unravel them. It is possible to note that instead of leading to a consideration of the differences between facts and values immersed in the decision-making process, CTNBio's control led to an intermingling of these elements through discourse and decisions based solely on scientific terms.

In this process, ethical and political considerations risk being marginalized as risk assessment by a "technical committee" will tend to reduce risk assessment to being a purely technical process. This can lead to a flawed regulatory process, as social, ethical, and political issues end up not finding any appropriate space for analysis and deliberation. This was evident in the analysis in the previous chapter when the distribution conflicts associated with GMOs were analyzed. Even if RR soybean could be considered environmentally safe, this does not mean that it cannot produce social and political effects of some kind that need to be considered in decision-making. It is the ethical and political nature present in the risks produced by new technologies that requires them to be examined by committees empowered to recognize these dimensions that characterize them.

It is possible that CTNBio's growing control over the decision-making process during this period led to a situation of institutional irresponsibility. To consider this hypothesis, one can take into perspective the warning given by Anvisa's president, Cláudio M. O. Henriques, during the period. With CTNBio's control over RA, he stated in the period, Anvisa "lost the power to refuse to grant such authorization for genetically modified products, should CTNBio have rendered a favorable opinion" (Henriques, 2008, s/p). The existing fear here is projected onto the possible loss of responsibility over decisions previously produced for GMOs. According to the president of Anvisa, the responsibility itself becomes volatile as it is "attributed to people with a two-year mandate, diluted in a commission that decides by voting" (Henriques, 2008, s/p). The problems to which this CTNBio control can refer us are indicated by the following analogy offered by him:

> [...] when a drug is found to produce unexpected harm, even if this occurs long after its registration, Anvisa is held responsible for having authorized its existence in the country and for all the necessary procedures to contain the risk, including, as the case may be, suspending production and recalling the product. (Henriques, 2008, s/p)

For the way the RR soybean decision was operated, this kind of accountability tends to be nullified. It is therefore possible to see that some of the minimal elements of a policy grounded in the PP have been nullified with CTNBio's technocratization process.

5.8 Final Considerations

Advocates for the commercial release of RR soybean offered a minimal precautionary approach to address the environmental risks entailed in the decision. This view basically relied on the application of an RA to guide the policy decision. Hence, its proponents did not see as necessary the use of additional precautionary measures to guide the established decision. In this perspective, there is no substantive rupture of the regulatory process as it is known since such procedures do little to change the measures already in place. The major difference in regulating GMO risk lies in the use of an RA that, given its characteristics, differs little from existing RAs. The critics (precaution alliance), in turn, have offered a set of arguments and proposals that go beyond the mere use of the RA to guide the policy decision. In this discourse, the release of RR soybean should operate so that its release is conditional on a broader set of precautionary measures. Differentiated studies should be adopted (EIA) and measures on consumption (labeling) and distribution (traceability) also incorporated.[10] At certain points, it is possible to see that these measures were part of an integrated perception of the regulatory process for GMOs.

The conflict over the environmental risks of RR soybean reproduces a type of cognitive dispute that tends to occur in debates over technological risks. A dispute that especially involves the use and interpretation of scientific knowledge to define these risks and, simultaneously, of making the authority conferred to knowledge end up influencing the policy decision itself. Therefore, it is possible to say that this axis of the conflict in Brazil ended up expressing a phenomenon that affects disputes over GMOs in other countries. The conflict over risk is translated as a type of dispute over its perception and, consequently, involves a clash of the different types of knowledge that make up the discussion on the commercialization of GMOs. If scientific authority is not the same as political authority, there is a close and almost direct relationship between the two in situations like these. The former tends to be seen as a direct passport to obtain the latter. Hence, the conflict over RR soybean release has not only placed critics of RR soybean release in opposition to the RA but also induced them to operate a distinction between RA and EIA. At different points in the conflict, these environmental policy instruments were seen as equivalent and at other times as distinct. A result of the dispute over the authority of knowledge that permeated the conflict. However, as already indicated, it is not certain that these environmental policy instruments express the scientificity that the actors in the conflict expect of them, much less create policy prescriptions that can be extracted almost automatically from the technical information they offer. Such assumptions tend to cloud rather than clarify the type of policy that should be adopted for the risks brought by new technologies incorporated into modern Brazilian agriculture.

References

Augusto, L. G. da S. (2005). *Integrante da CTNBio se desliga após liberação*. Disponível em: https://fase.org.br/pt/informe-se/noticias/integrante-da-ctnbio-se-desliga-apos-liberacao/. Acesso em: 19 ago. 2020.

---

[10] Some of the elements that were present in the dispute over possible preventive measures that should be adopted for RR soybean release will return in part in the next chapter. As will become evident, for critics of RR soybean release, specific labeling for GM foods will not be seen merely as an information tool for the consumer, but also as a precautionary measure associated with the risks and uncertainties that are generally linked to GMOs. This measure would, in turn, be linked to the proposals on traceability and segregation of GM crops.

Beattie, R. B. (1995). Everything you already know about EIA (but don't often admit). *EIA Review, 15*(2), 109–114.
Bresser-Pereira, L. C. (1999a). *Alimentos transgênicos e biossegurança*. Disponível em: http://www.bresserpereira.org.br/view.asp?cod=596. Acesso em: 19 ago. 2020.
Bresser-Pereira, L. C. (1999b). In defence of science. In *World conference on science, June 28–July 1, 1999*, UNESCO. Disponível em: https://www.bresserpereira.org.br/documents/MCT/99-91Budapest-MCT.pdf. Acesso em: 19 ago. 2020.
Bresser-Pereira, L. C. (1999c). *Audiência pública realizada pela comissão de defesa do consumidor, meio ambiente e minorias. No. 0533/99. Debate sobre a autorização para produção e consumo de alimentos transgênicos no país*. Disponível em: https://www2.camara.leg.br/atividade-legislativa/comissoes/comissoes-permanentes/cdc/documentos/notas-taquigraficas/not.1999.html/nt16061999.pdf. Acesso em: 19 ago. 2020.
Brunner, E., Millstone, E., & Mayer, S. (1999). Beyond 'substantial equivalence'. *Nature, 40*, 525–526.
Cesarino, L. M. C. N. (2006). *Acendendo as luzes da ciência para iluminar as luzes do progresso*. Dissertação (Mestrado em Antropologia Social) – Programa de Pós-Graduação em Antropologia Social. Universidade de Brasília.
Cezar, F. G. (2003). *Previsões sobre tecnologias: pressupostos epistemológicos na análise de risco da soja transgênica*. Dissertação (Mestrado em Filosofia) – Departamento de Filosofia, Universidade de Brasília.
Colli, W. (2004, 8 abr). *Transgênico é palavrão? Boletim Economia Política*.
Conama. (2002). *Conselho Nacional de Meio Ambiente. Resolução Conama n° 305, de 12 de junho de 2002*. Publicada no DOU no 127, de 4 de julho de 2002, Seção 1, páginas 81–82. Dispõe sobre Licenciamento Ambiental, Estudo de Impacto Ambiental e Relatório de Impacto no Meio Ambiente de atividades e empreendimentos com Organismos Geneticamente Modificados e seus derivados. Disponível em: https://www.mma.gov.br/estruturas/biosseguranca/_arquivos/71_01122008102705.pdf. Acesso em: 13 ago. 2020.
Cranor, C. F. (1993). *Regulating toxic substances. A philosophy of science and the law*. Oxford University Press.
de Castro, L. A. B. (1998). Biossegurança: realidade e perspectiva no Brasil. *Biotecnologia, Ciência e Desenvolvimento, 1*(6), 4–8.
de Castro, L A. B. (2008). *A Ciência no Brasil. Camões e os Transgênicos*. Disponível em: http://www.alerta.inf.br/índex.php?news=265. Acesso em: 10 jul. 2008.
Divergência faz técnica da Unicamp deixar CTNBio. (1999). *Jornal Folha de São Paulo, 22 de junho de 1999*. Disponível em: http://www1.folha.uol.com.br/fsp/agrofolh/fa22069916.htm. Acesso em: 4 abr. 2008.
Fischoff, B., et al. (1981). *Acceptable risk*. Cambridge University Press.
Greenpeace. (2006). *O Princípio de Precaução e os Transgênicos*. Disponível em: http://www.pick-upau.org.br/mundo/transgenicos_agora/02_principio_precaucao.pdf. Acesso em: 10 ago. 2019.
Greenpeace. (2008). *Consumo de herbicida aumenta com uso de plantas transgênicas*. 2004. Disponível em: http://www.greenpeace.org/brasil/documentos/transgenicos/greenpeacebr_040517_transgenicos_documento_super_ervas_port_v1. Acesso em: 07 jul. 2008.
Henriques, C. M. P. (2008). *A CTNBio e a expropriação da regulação*. Jornal do Brasil, 4 mar. 2007. Disponível em: http://www.jornaldaciencia.org.br/Detalhe.jsp?=45034. Acesso em: 10 ago. 2008.
IDEC. (2003a). *Soja transgênica: Idec entra na justiça contra a União Federal e o Estado do Rio Grande do Sul*. Publicado em 02/10/2003. Disponível em: http://www.idec.org.br/emacao.asp?id=484. Acesso em: 15 jul. 2008.
IDEC. (2003b). *Ação Civil Pública com Pedido de Tutela Antecipada. Ação Civil Pública do Idec contra a União Federal e o Estado do Rio Grande do Sul*. Publicado em 02/10/2003. Disponível em: https://trf-1.jusbrasil.com.br/jurisprudencia/990581/apelacao-civel-ac-34026-df-20033400034026-7/inteiro-teor-100612129?ref=juris-tabs. Acesso em: 19 de ago. 2020.

References

IDEC. (2008). *Equívocos e omissões de Veja sobre os transgênicos*. 28 out. 2003. http://www.idec.org.br/emacao.asp?id=499. Acesso em: 22 ago. 2008.
Justiça Federal. (2008). *Seção Judiciária do Distrito Federal. Ação Cautelar Inominada Decisão No. 99, processo No. 1998.34.00.027681-8*. Requerente: Idec, Requeridos: União Federal e Outro, Elaborado por Antonio S. Prudente. Jus Navegandi, n.33. 1998. Disponível em: www1.jus.com.br/pecas/texto.asp?id=335. Acesso em: 10 jul. 2008.
Lacey, H. (2006). *A Controvérsia sobre os Transgênicos*. Ideias e Letras.
Lajolo, F. M. (2004). Alimentos transgênicos: riscos e benefícios. *Ciência Hoje, 34*(203), 36–37.
Lexer, W. et al. (2008). *Risk assessment. D3.2 Report WP 3*. Disponível em: www.umweltbundesamt.at/fileadmin/site/umweltthemen/UVP_SUP_EMAS/IMP/IMP3-Risk_Assessment.pdf. Acesso em: 14 ago. 2008.
Lohani, B. N. et al. (1997). *EIA for developing countries in Asia*, v. 1, Overview. Asian Development Bank. Disponível em: http://www.adb.org/documents/Books/Environment_Impact/. Acesso em: 27 jul. 2008.
Lutzenberger, J. (1999). *Soja transgênica. Problema Político, Não Técnico*. Disponível em: http://www.Fgaia.org.br/texts/t-transgenicos.html. Acesso em: 27 jul. 2008.
Marinho, C. L. (2003). *O discurso polissêmico sobre plantas transgênicas no Brasil: estado da arte*. Tese (Doutorado em Saúde Pública) – Escola Nacional de Saúde Pública.
McGarvin, M. (2001). Science, precaution, facts and values. In T. O'Riordan et al. (Eds.), *Reinterpreting the precautionary principle*. Cameron.
Mooney, C. (2006). *The republican war on science*. Basic Books.
Myhr, A. I. (2007). Uncertainty and precaution: Challenges and implications for science and the policy of genetically modified organisms. In N. de Sadeleer (Ed.), *Implementing the precautionary principle. Approaches from the Nordic countries, EU and USA*. Earthscan.
Myhr, A. I., & Traavik, T. (2002). The precautionary principle: Scientific uncertainty and omitted research in the context of GMO use and release. *Journal of Agricultural and Environmental Ethics, 15*(1), 73–86.
National Research Council. (1994). *Science and judgment in risk assessment*. National Academy Press.
Pavan, C. (2008). *Está provado cientificamente: os OGMs são mesmo seguros. Entrevista com Crodowaldo Pavan*. Conselho de Informações sobre Biotecnologia. Disponível em: http://www.cib.org.br/entrevista.php?id=22. Acesso em: 20 jul. 2008.
Pessanha, L., & Wilkinson, J. (2005). *Transgênicos, recursos genéticos e segurança alimentar. O que está em jogo nos debates?* Armazém do Ipê.
Porto, M. (2002). *A batalha jurídica ainda não terminou*. Disponível em: http://www.comciencia.br/dossies-1-72/reportagens/transgenicos/trans03.htm. Acesso em: 15 fev. 2009.
Raupp, M. A. (2008). *Uma decisão de grande responsabilidade*. Jornal do Estado de São Paulo. Disponível em: http://www.estado.com.br/editorias/2008/02/10/opi-1.93.29.20080210.2.1.xml. Acesso em: 20 mar. 2009.
Sadeleer, N. (2002). *Environmental principles*. Oxford University Press.
Sadeleer, N. (2007). The precautionary principle in European Community health and environmental law: Sword or shield for the Nordic countries? In N. de Sadeleer (Ed.), *Implementing the precautionary principle*. Earthscan.
Scholze, S. H. C. Biossegurança e alimentos transgênicos. *Biotecnologia, Ciência e Desenvolvimento*, Brasília, v. 2, n. 9, p. 32–34, jul.-ago. 1999.
Transgênicos e agrotóxicos. (2001). *Biotecnologia, Ciência & Desenvolvimento, 3*(18), 51.
van den Belt, H., & Gremmen, B. (2002). Between precautionary principle and "sound science": distributing the burdens of proof. *Journal of Agricultural and Environmental Ethics, 15*(1), 103–122.
von Schomberg, R. (2006). The precautionary principle and its normative challenges. In E. Fishcer et al. (Eds.), *Implementing the precautionary principle. Perspectives and prospects*. Edward Elgar Publishing Limited.

Chapter 6

Labeling as Precaution: Substantial Equivalence and the Conflict over Labeling

> When I presented this project [of labeling], I did not intend to define all the politics regarding transgenics in Brazil (…) Evidently, it was not only a question of consumer choice, I based myself on the precautionary principle. (Representative Fernando Gabeira)
>
> The government has not decided to label because it thinks it is dangerous. The government has decided to label because it thinks it is a consumer's right to know if the product he or she is consuming is a transgenic product. (Bresser Pereira, former Minister of CAT)
>
> Over the last six or seven years, when regulating transgenics, the governments, and even the scientific community, all the time, committed themselves to the belief that the consumer would be assured the right to information, so that he or she could then exercise his or her legitimate right to eat or not eat. All these promises, as we know, have not been kept. The substantive truth is that, after all, there has never been, until today, any labeling on products of transgenic origin in Brazil. (Luiz Eduardo R. de Cawalho, former president of the Brazilian Society of Food Science and Technology)

In this chapter, the conflicts surrounding the labeling of RR soybean will be examined. The conflict will be examined from what will be referred to as the labeling story line. The latter represents the axis through which the various structuring issues of the labeling conflict presented themselves in the Brazilian case. Among the points that will be examined in this story line is the view on the relationship between labeling, precaution, and consumer choice. From this analysis, it is possible to verify two distinct discourses on labeling. On one side is the discourse of precautionary labeling and on the other, the discourse of conventional labeling. These discourses present different assumptions about labeling and its relation to issues associated with Principle of substantial equivalence (PSE), science, risk, and nutritional safety (see Table 6.1). The remainder of the chapter is devoted to examining these differences in the conflict over labeling in Brazil.

Table 6.1 The labeling dispute

Structuring issue (*issue framing*)	Conventional labeling alliance	Precautionary labeling alliance
Market	Specific labeling implies market loss for Brazilian agricultural products.	Lack of specific labeling can mean loss of market share, especially for those countries that are introducing labeling policies like the EU.
Precautionary principle and labeling	Labeling for products containing GMOS is decoupled from precaution. It has a purely informative character for the consumer.	Critique of the Sound Science approach to labeling. Specific labeling of product containing GMOs is tied to precaution. It allows for continuous safety assessment of GM products.
Science	Sound Science approach to labeling. The information that labeling must contain is restricted to scientific information about the nutritional content of foods. Specific labeling lacks scientific justification, as products are considered to be equivalent. Distrust of science to report levels of transgenicity.	Specific labeling is based on a political decision to defend the consumer's right to know whether or not a food is GM. But specific labeling is also a scientific decision, because it allows the study and permanent monitoring of the risks posed by GM products. Reliance on science to detect levels of transgenicity in GM products.
Substantial equivalence	The existence of substantial equivalence between RR soybean and conventional soybean makes specific labeling for GMOs incoherent and unnecessary.	Rejection of the principle of substantial equivalence. GM and conventional products are seen as not equivalent, requiring specific labeling for GM products.
Risk and safety	The risks posed by RR soybean are the same as for conventional soybean. Labeling is disconnected from risk and safety issues related to GMOs.	The risks of RR soybean are not yet fully known and the risks posed by RR soybean cannot be seen as equivalent to the risks attached to conventional soybean. Specif labeling is closely linked to food safety and environmental issues.
Risk analysis (RA)	By demonstrating the safety of RR soybean, the RA is used as a means to justify conventional labeling for RR soybean.	The RA is not an effective means of assessing the risks posed by RR soybean and therefore does not justify the rejection of specific labeling for GM products.
Positive and negative labeling	Presenting arguments that disqualify the need for specific positive and negative labeling for GMOs.	Specific and Positive labeling ("contains GMOs") as mandatory.
Voluntary and compulsory	Disqualification from specific and mandatory labeling for GMOs.	Specific and Mandatory labeling requirement for GMOs.
Product/process	Emphasis on the product and its nutritional information. Evaluation through nutritional components.	Emphasis on process. Concern with the process by which the product is made. Evaluation through the type of productive system (existence of transgenic).

(continued)

Table 6.1 (continued)

Structuring issue (*issue framing*)	Conventional labeling alliance	Precautionary labeling alliance
Tolerance	Comprehensive tolerance levels. Exclusion of foods that cannot be identified. For included foods, the tolerance level is always higher than the GMOs' critics propose.	Full labeling: inclusion of all foods that undergo some genetic modification. Requirement of minimum tolerance levels (this does not exclude labeling of all products containing GMOs, which would be possible with measures involving traceability).
Law	Consumer legislation justifies the conventional labeling of GM products. The absence of specific labeling does not disrespect the consumer's right.	Consumer legislation justifies the requirement for specific labeling for GM products. Conventional labeling does not respect consumer rights.
Paternalism	The release of RR soybean is founded on CTNBio's authority. Knowledge deficit between CTNBio experts and lay people. The rejection of GMOs is due to the ignorance of critics.	CTNBio's decision disrespects the biosafety law. Refusal of the paternalism conferred on the CTNBio. The consumer has the autonomy to make his or her choices. The refusal is not due to ignorance or lack of knowledge, but for legitimate reasons.
Ideology	Specific labeling leads to discrimination against GM products. It involves an arbitrary distinction between "substantially equivalent" products. Therefore, labeling can involve manipulation of the consumer by generating "fears" and "mistrust" without scientific basis.	The absence of specific labeling for GMOs involves consumer manipulation. People who do not wish to consume GM products can consume these products without being aware of this fact.
Consumer rationality	The consumer is seen as a passive economic agent. They do not have the appropriate cognitive conditions to choose between GM and conventional products. Product cost considerations predominate. To the extent that labeling makes products more expensive for reasons that are seen as illegitimate, it is seen as promoting consumer irrationality.	The consumer presents himself as a "Homo economicus." Their choices can influence the economic process and technological innovation. It is up to the consumer to decide whether GMOs are desirable or not. Specific labeling allows us to respect the consumer's different rationalities and the ethical, cultural, and economic values that guide their choice.
Autonomy	The choice between GM and conventional products is considered irrelevant to realizing consumer autonomy. This choice should be left to the experts and regulatory bodies.	The choice between GM and conventional products is a condition for consumer autonomy. Specific labeling is a condition for consumers to be able to associate values and life choices with their decisions through the consumption process.
Responsibility	It is not up to CTNBio to make decisions regarding labeling. Specific labeling is a political issue, not a technical one.	The responsibility lies with CTNBio and other regulatory agencies. Specific labeling is a political and technical issue.

Source: The author

6.1 Labeling Risks

In the last decade, many countries have implemented policies aimed at labeling GM foods. And beyond the label itself, the conflicts and tensions that such a policy usually engenders have also become common. For companies, the label is a central element of product marketing policy and thus tends to be seen as directly influencing consumer decisions. For environmentalists, on the other hand, the label tends to be seen as a space of symbolic struggle and as a means of achieving better regulation of these products. For these reasons, it seems natural that labeling constitutes an important space for struggles that seek to define the commercialization of GMOs.

Unlike the discussions surrounding the use of the RA in the release of RR soybean, the debate over labeling would seem to offer a more consensual framework for the implementation of a policy on labeling GM foods in the country. After all, from the beginning, CTNBio and government representatives were in favor, at least in discourse, of labeling these products. In 1999, the former minister of Science and Technology, Bresser-Pereira, informed in a public hearing, for example, that: "Regarding the problem of labeling [...] the government decides for labeling and there is no problem of time, of time inconsistency, because what the government authorized now was the planting of seeds" (Bresser-Pereira, 1999b, p. 11). But if there was an apparent consensus to label GM products, why has labeling fomented such intense conflict between the government and civil society groups, as will be seen below?

First, it is necessary to consider that since the beginning of CTNBio's operation, civil society organizations have not seen any serious commitment from the biosafety commission to a labeling policy. In 1996, IDEC sent a series of requests to the commission, including the creation of a labeling policy, without getting any response on the issue, which will cause Greenpeace to disassociate itself from the commission as early as 1997. At least, this was one of the justifications for its withdrawal from the committee in that period. This disinterest was based on the very limits of responsibility that the members of the CTNBio attributed to the commission in its role of implementing a labeling policy for GMOs. According to Esper Cavalheiro (2001, s/p), former president of CTNBio: "The labeling issue transcends the legal powers of CTNBio, since it involves issues related to consumer protection and therefore concerns the consumer defense code." In the same vein, Leila Oda (2000c, s/p), also former president of the commission, argued in the period that labeling is a "much more political issue than technical". With this, she sought to indicate that, with CTNBio being a technical commission, it would not be its responsibility to make decisions associated with labeling. Which, according to this argument, would present itself as an eminently political decision. Which helps us understand why labeling was not even a topic of analysis in the debates when the CTNBio was created. Something that was absent in the process of commercial release of GMOs.

On the commercial release of RR soybean, the government's position will present some additional inconsistencies. The release of RR soybean, it seems, would

6.1 Labeling Risks

lead to a gradual and automatic commercial release of the product. After all, the approval occurred without the government coming up with any specific labeling program for GMOs. This provoked an immediate reaction in civil society organizations. For this reason, in the public civil action filed by IDEC against the CTNBio in the period, the organization claims that, before analyzing and issuing a conclusive technical opinion to Monsanto's request, the government should regulate the "food safety, commercialization and labeling norms. Without which it cannot evaluate any application" (IDEC, 2005, 2008b, s/p). Or, as will be observed by Congressman Fernando Gabeira in the period of the release:

> Why does the government decide to release and, simultaneously, label, but regulate labeling later? Well, if the Brazilian Government considers that liberation and labeling are interconnected, that labeling is important for the preservation of consumer health, or at least for their ability to choose, it cannot liberate before defining all the processes linked to labeling, otherwise it is a contradiction. (Gabeira, 1999, p. 9)

It might be assumed that the commercial release for planting would await additional studies to ascertain the safety of RR soybean. However, this assumption has been contradicted by CTNBio and the government, which have come to view the RA as a satisfactory and sufficient means of examining the risks posed by RR soybean. The report presented by Monsanto, and endorsed by CTNBio, to justify the commercial release, states, for example, that "the results of technical and scientific studies carried out with RR soybean in Brazil in other countries demonstrate its environmental safety" and that part of these conclusions can be drawn from the "substantial equivalence of the RR soybean in relation to the non-genetically modified soybean" (Monsanto, 1998 apud Cezar, 2003, p. 79). Leila Macedo Oda, former president of CTNBio, will also state in 2001 that "labeling is information" and that "the product to be labeled has already passed all the safety tests," being labeling a "much more political issue than a technical one" (Oda, 2000c, s/p). But this argument seems to contain a contradiction within itself. For if CTNBio's "technical" decision would serve to sanction the commercial release of RR soybean, then this same "technical" decision could also be seen as a "political" decision. Not least because there would not be, after CTNBio has made its decision, any government body to examine the issue in political terms. Thus, the question that this discourse seemed to pose in the period was: Considering that the safety of RR soybean had already been proven by the RA, what would be the impediment to release it commercially immediately? In this context, it seems clear that commercial release would be done automatically and following the usual labeling processes.

The importation of GM soybean from Argentina, which will occur during the period, will prove to be another important point for the unfolding of the conflict. This fact will cast a veil of doubt over the real intentions of the government and its ability to conduct the regulation of GMOs in the country. It should be remembered that MS&T and CTNBio had not taken any action to reverse the occurrence and did not even try to communicate their opposition to the regulatory policy that was being implemented. This suggests that, for the government, there was no impediment to the commercialization of RR soybean at that time. It had even allowed the importation of GM seeds. At the same time, this ambivalence of the government towards labeling was accompanied by an enthusiastic discourse on the role of agbiotech for the country's scientific and economic progress, a discourse that contributed to the perception by environmental groups that economic issues were taking precedence over issues of environmental safety and respect for the consumer.

6.2 Defending the Consumer's Right in the Absence of Danger

In the conflict over labeling of GM food, the alliance of conventional labeling presented a position that can be found in other countries. One can take this into account to better understand the position of the proponents of the commercial release of RR soybean in the conflict. In the United States, for example, the Biotechnology Industry Organization (BIO) came out in favor of consumer rights, but at the same time came out against specific and mandatory labeling on the grounds that it could end up confusing consumers (Klintman, 2002). In Brazil, something similar occurred. It was not uncommon for members of the government, CTNBio, and the food industry to come out in favor of consumer rights, but it was also not uncommon for them to present arguments that sought to invalidate the implementation of specific labeling for GM products. The fact that the government, CTNBio, and the industry were, in the conflict, in favor of labeling does not imply that their representatives were, in fact, in favor of mandatory and specific labeling for these products. This is because, as will be shown later, indicating that one is in favor of labeling GMO products may simply suggest that one is in favor of conventional—and not necessarily specific—labeling for these products. In this case, although GM foods may be labeled, such an option would not allow for the differentiation of GM products from conventional products, since both would be labeled to the same standard. In order to analyze this ambivalence in the discourse of these actors, the different arguments that are usually put forward to invalidate the labeling of GM foods need to be examined and then put into perspective the Brazilian case.

The economic costs associated with labeling are usually presented as a way to refuse a specific labeling system for these products. Since labeling can make GMOs more expensive, the economic benefits associated with them could be lost to investors and consumers. This could especially be the case if the costs were sufficient to make it impossible to offer these products on the market. Another argument is that specific labeling leads to an arbitrary distinction between "substantially equivalent" products. This argument unfolds along legal lines, arguing that since GM products and conventional products have equivalent nutritional properties, there would be no moral and legal grounds even to justify labeling that discriminates between one type of product and another. Therefore, since there is no scientific and legal basis to differentiate GM products from conventional ones, labeling could be accused of generating a kind of guilt or stigma without any rational justification. The GM food

6.2 Defending the Consumer's Right in the Absence of Danger

would be considered as more dangerous when, in fact, they would be equivalent to the others. This would occur when, for example, a label would indicate that a GM product is more dangerous than its conventional equivalent, even though, according to the defenders of the release of RR soybean, there is no scientific basis for making such an association.

It is also argued that labeling tends to be misleading and irrelevant even when its information appears to us to be correct. For example, an American scientist opposed to labeling has gone so far as to say that "even a message that is accurate, in the narrowest sense, can mislead and confuse consumers if it is irrelevant, unintelligible, or so craftily selected that it provides inadequate or biased information" (Miller, 1995 apud Klintman, 2002). In other words, although GM products are different from conventional products, which in this case would justify specific labeling for them, such a measure would induce consumers to perceive them as being more dangerous than they really are. Finally, the proponents of conventional labeling for GM foods also rely on the impossibility of assessing the level of transgenicity, pointing out the technical impossibilities of achieving this goal. Since this level may prove impossible to assess, any attempt to discriminate GM products from others would prove unfeasible or useless to adopt.

Despite speaking out in favor of consumer rights, it was not uncommon for representatives of the government, CTNBio, and the food industry to help promote some of the very arguments that are used to disqualify specific labeling for GM foods elsewhere. Bresser-Pereira's and CTNBio's thesis that RR soybean was a substantially equivalent product, for example, forms the basis of the FDA's discourse in the United States against labeling these products. Although the substantial equivalence thesis was used in the debates about the environmental risks of GMOs, the thesis was also used in the labeling debates, since this discourse tends to reinforce the prescription of conventional labeling for these products. In the Brazilian conflict, there were not a few moments in which arguments to discredit specific labeling to GMOs were presented. Lynn Silver, for example, IDEC representative, reports the following dialogue she had with the president of CTNBio during the liberation period:

> Actually, what I understood from Dr. Barreto de Castro's speech was not clear whether it was his personal position or the commission's position, but that, if the risks to human health were not identified, the inclusion of the indication of genetic engineering on the label was irrelevant. (Silver, 1999, s/p)

In 1999, after the court decreed the need for a labeling policy of GM products in the country, several arguments against labeling were again raised in public hearings in the country. It was expressed, for example, by Congressman Luciano Pizzato, who indicated that, in the absence of scientific facts about the risks of RR soybean, specific labeling for GMOs would prove incoherent. According to him:

> we are here discussing labeling and not the ideological issue of whether or not we want transgenics, what are you going to warn? Which product is transgenic and can be bad for your health? We cannot even mention this, there is not the slightest indication of this. (Pizzato, 2000, p. 17)

It is not uncommon for this perception on the impossibility of examining the risks of GMOs ("they are going to warn what?") to be accompanied by a recognition of the right of the consumer, as the same deputy does, which is presented as follows: "I believe it is a pacific point in this committee the view that, if it exists, the trade of transgenics [...] one should warn with proper labeling" (Pizzato, 2000, p. 15). At first glance, this might seem contradictory. After all, if a specific labeling to GM foods seems to be unfounded, how would it be possible to defend the consumer's right in such a scenario? The recognition of the impossibility of specific labeling for GM foods does not, therefore, prevent the defense of conventional labeling for them. One type of labeling (specific) is criticized through the defense of another (conventional). In this way, although the refusal of labeling is not always evident in the liberation discourse, the criticism of its irrationality is always a present aspect of this discourse. Specific labeling for GM foods tends to be seen as irrational, unnecessary, and misguided. Advocates of the release of RR soybean point to the informational inconsistency of a differentiated labeling system and its limited validity in fostering rational consumer choice. Hence, conventional labeling for these products tends to be perceived as the solution to the dilemma raised by the RR soybean trade.

The informational inconsistency of a specific labeling for GM products has been justified with different arguments. In the first, it would be based on the arbitrary classification that labeling establishes among substantially equivalent products. In public hearings on the issue, according to Congressman Confúcio Moura (2001), the labeling project presented by Congressman Gabeira in 1999, which provided for the labeling of GM products, proved to be controversial "because countries, particularly the United States, consider the placement of the specific label to be discriminatory." The labeling would also be inconsistent due to the "difficulties in indicating the ingredients and by-products in the composition of the food." Therefore, for the deputy, "labeling would be a very difficult and discriminatory way for certain products" (Moura, 2001, p. 12). It is possible to see in these passages how the Principle of substantial equivalence (PSE) was used in this speech in order to invalidate the need for labeling. When the former president of CTNBio, Dr. Barreto de Castro, informs that if the risks of soybean were not identified, the demand for a specific labeling for it would be irrelevant; he is also using the Principle of substantial equivalence (PSE) in his discourse, since the risks of soybean were examined based on the premises of this principle. So will Bresser-Pereira (1999b, s/p), former Minister of Science and Technology, when he says that, in the case of soybean, "there is no substantial change in the product" and that, for this reason, one can conclude that the "product is exactly the same." Since conventional RAs are sufficient and satisfactory to assess the risks posed by these products, the prescription of conventional labeling systems follows as a logical development of the argument, although this logical connection is not stated. Like the congressman seen above, former Minister of Science and Technology Bresser-Pereira will state this while arguing that the government would follow the European labeling policy, in which "you indicate, whenever necessary, that the product contains genetically modified product" (Bresser-Pereira, 1999b, p. 6). However, as will be seen later, this statement contains a contradiction when examined in the context through which the European

6.2 Defending the Consumer's Right in the Absence of Danger

GMO labeling policy has developed. This is because even European policy was not clear about the labeling process for GM foods at the time, which makes this clarification by the minister end up clarifying very little. Hence, for this and other reasons, a moratorium on GMOs in Europe would be launched in 1999, the year that RR soybean was just being released in the country.

The Principle of substantial equivalence (PSE) played a central role in the conflict over the labeling of GM foods. The presence of this principle in government discourse demonstrates that the premise that CTNBio used to guide the decision for the commercial release of RR soybean had direct implications for the debate over its labeling in the process. It is this premise that in some way supports the accusations of discrimination made by parliamentarians in the public hearings. This is because discrimination is based on the accusation that labeling is making things different that are in fact the same or equivalent. And this accusation will occur in different ways. At first, it is perceived as existing in the comparison that will be made with other GMOs, but not agricultural products. So, returning once again to the speech of Congressman Luciano Pizzato: "We have genetically modified trees. Why do we only discuss agricultural products? Why hasn't the Department [...] demanded yet that genetically modified drugs warn in the labeling that they are transgenic? Why is this commission forgetting this?" (Pizzato, 2000, p. 16).

A second type of discrimination is seen as being associated with the type of communication provided by the labeling itself. The food industry's refusal to incorporate specific labeling for GM products into the conflict was justified by the possible distorted communication that the labeling would produce. Abia's legal director will state in the period, then, that: "[the labeling requirement] is in force, but it is not incorporated, because the industry does not want to attach its brand to a warning, as if it were a dangerous thing" (Abia's Legal Director, 20--? apud IDEC, 2008c). This concern is present in the draft of Legislative Decree No. 90 of 2007, by Congresswoman Kátia Abreu, which states that the label to be placed on products "refers to the idea of attention and care and may encourage distrust of the population in products that have already been evaluated and considered safe by the National Technical Commission on Biosafety (CTNBio), thus hindering the introduction of these products on the market" (Abreu, 2007 apud BRASIL, 2007, s/p). The argument shows once again how CTNBio's position continued to have a strong influence in invalidating the specific labeling for GM foods. The latter had become unnecessary because the RR soybean had been "evaluated and considered safe by the National Biosafety Technical Commission" (Abreu, 2007 apud BRASIL, 2007, s/p). The argument by former CTNBio president Leila Macedo Oda that labeling is a "much more political than technical issue" conceals the fact that the scientific considerations used by CTNBio to release soybeans have been used repeatedly to disqualify the need for labeling GM products in the country. The political debate over labeling was at no point dissociated from the scientific controversies surrounding its commercial release. The scientific issues (the validation of the release of GMOs through an RA) were seen as sufficient to inform the policy decisions (the labeling and release).

Some of the claims about discrimination, however, do not say where the communicative distortion they say exists. Some arguments suggest that no matter how the labeling is presented, it will always unequivocally induce consumer irrationality. These criticisms do not always contest whether the message offered by the labeling is true or false. They simply challenge the labeling for the simple irrational effect it could produce. Thus, the fear and distrust that it could generate in consumers are perceived as having no valid scientific basis for consumer decision-making.

6.3 Consumer Choice and Environmental Risk

There are several reasons why people might be interested in the labeling of GMOs. The main one is consumer freedom of choice. Labeling can allow people to make choices about cultural prohibitions (vegetarianism and animal welfare) and health risks in order to link their consumption choices to their lifestyles. In this case, by offering a range of information linked to consumer values, labeling can serve as a means to enable consumers to make autonomous choices. The precautionary labeling alliance has relied on arguments close to these to justify the specific labeling of GM foods. It should be noted that labeling was first seen as involving a type of precaution. The link between labeling and precaution will appear in two different ways in the discourse of the precautionary labeling alliance. One occurs for a strategic reason when this discourse mentions the use of labeling as a way to delay the commercial release of RR soybean in order to promote a precautionary measure with this type of action. In 2001, when Gabeira's labeling project served as the basis for an initial discussion about labeling in the country, the deputy said that "it was not just about consumer choice, it was based on the precautionary principle. I thought it was necessary to delay a little the process of entry of transgenics into Brazil" (Gabeira, 2001, p. 2). Thus, Gabeira's labeling project, which began to guide the initial debate on labeling in 1999, was directly associated with this principle. Judge Antonio Souza Prudente's decision in the civil suit brought by IDEC and Greenpeace will reveal a similar view in which precaution and labeling are also seen as interconnected. In his decision, the judge will note that the: "mere labeling of transgenic products seems insufficient to fulfill the effectiveness of the prevention principle [...]" (Prudente, 2000, s/p). Although considered insufficient, labeling is seen here as an integral component of the principle. The judge mentions the limits of labeling to apply the precautionary principle because he seeks precisely to establish a correlation between labeling and environmental impact studies (EIS) as a means of applying precaution in more general terms. Along the same lines, Marilena Lazzarini, a lawyer with IDEC, reports that, as far as labeling is concerned: "there is a need to adopt a precautionary stance in this process, aiming to ensure consumer – and environmental – protection in relation to transgenic foods" (IDEC, 2008a). And Lynn Silver, from the same organization, will report that "labeling is indispensable for the identification, in the future, of adverse effects that may arise after the introduction of the products." For her: "besides being a consumer

6.3 Consumer Choice and Environmental Risk

right, labeling is an essential tool for the control of unexpected effects from the point of view of human or even animal health" (Silver, 2001, p. 26).

The relationship between labeling and precaution is not always recognized, given the different ways in which the two can be linked, which would make labeling an integral part of precaution. Rather than an absolute ban, precaution can be understood as enabling a wide range of measures that do not always result in a ban on marketing a product. Precaution, as Whiteside (2006) points out, may imply the need for ongoing, or long-term, vigilance, since in many cases it is not possible to reach an immediate decision about the harmlessness of products. And, in the case of GM foods, labeling as a precautionary measure may be necessary, since it allows investigating whether the monitored activities are responsible for unexpected effects that may arise. It was precisely the absence of a consistent labeling policy that was the basis for European countries declaring a moratorium on GMOs in the late 1990s. Soon, the most significant contemporary measures in the field of labeling of GM foods, imposing mandatory labeling for all such products in the EU, came at the moment when it became recognized that precaution required labeling and traceability of GM foods (Whiteside, 2006, p. 24). The decision to make room for remedial measures in the future is at the heart of precautionary policy. The view of labeling as a means of identifying unexpected effects is also the main feature of labeling policy in the EU for GM products. And it is this sense which, as seen earlier, has been attributed to labeling by organizations such as IDEC.

The precautionary labeling alliance also relied on a legal argument and a moral argument to support specific labeling of GMOs. The legal argument is based on the thesis that consumer legislation would be in favor of specif labeling to GM food Judge Souza Prudente's decision, which sought to annul the decision to commercially release RR soybean, was to some extent based on this argument. In the civil suit filed by IDEC, the judge will state that:

> If it is indisputable that, according to article 6, II and III, the consumer has the basic right to adequate and clear information, with the correct specification of characteristics, composition, quality, and risks they present, among other data, it is also certain that only this data will provide the consumer's adequate right to choose, also assured by the Consumer Defense Code. (Prudente, 2000, s/p)

Organizations such as Greenpeace and IDEC also saw these rights as a basic legal condition to justify the labeling of RR soybean. However, this legal interpretation was challenged by a subsequent court decision. It was challenged by the decision of Judge Selene Maria de Almeida on February 25, 2002, who granted an injunction authorizing the planting and marketing of Roundup Ready (RR) soybeans. The judge's decision suspended the decision of Judge Souza Prudente and thus contradicted the interpretation that the consumer code would be sufficient to require specific labeling of these products. The judge's decision is emblematic, since she breaks with the usual interpretation that informs that the consumer legislation would provide the legal basis to require the specific labeling of GM products. In a report presented by the National Biosafety Association (ANBio), it is reported that "Judge Selene's report makes it clear that there are no reasons of technical-scientific or legal nature that prevent the commercialization of RR soybeans in Brazil approved by CTNBio's communiqué 54" (ANBio, 2008, s/p). The judge's decision suspended the requirement for a specific labeling for GM foods, creating the conditions for them to be marketed through conventional labeling in which there would be no indication of GMOs in the products.

While the requirement to conduct EIA was supported by environmental legislation, the labeling requirement was supported, in turn, by consumer legislation. This indicates that while the environmental legislation is clearer on the requirement to carry out EIA for innovations that may result in some type of impact on the environment, it is much less precise on the labeling requirement for GMOs. In this process, the labeling requirement has been left to the interpretation of the consumer code and how it can be interpreted to require specific labeling for GM foods. While labeling advocates saw consumer legislation as a strong basis for requiring labeling, representatives of the food industry interpreted that legislation in their favor and in a different way. The reason for this contradiction appears to lie in the fact that while consumer legislation appears to provide a justification for labeling, this same legislation was not produced to address the challenges associated with GM products, favoring a labeling standard for these types of products. It provides a similar legal standard for conventional products and GM products in particular. Critics might argue that since the legislation provides a legal standard that justifies labeling for both types of products, then this would justify labeling GMOs. But this argument tends to overlook a fundamental point of the issue. Advocates of the commercial release of RR soybean have not advocated not labeling GM foods, as seen in the previous part. What they, in fact, advocated is that GM foods be subject to the same labeling as other products. The non-labeling in the conflict represents just that: the use of conventional labeling. To the extent that GM and conventional products are seen as equivalent in terms of risk, what proponents of the commercial release of RR soybean advocated in the conflict was that GM products should then be labeled as conventional products. It was argued that consumer legislation should then be interpreted equally for both conventional and GM products, since these products could be classified as substantially equivalent. Which implied the use of a single labeling standard—the conventional one—for the products. It also implied that it was impossible to differentiate GM products, through labeling, from other products. To assume a differentiated policy for GM foods would mean, therefore, to adopt a discriminatory policy for these products.

Consider, for example, the following statement by the former president of CTNBio, Leila Oda, in the period. According to her: "We will not label to say that the product is dangerous, but to respect the consumer's right to choose, whatever the reason. If the person is allergic to eggs, he or she has the right to know that a final product, transgenic or not, contains egg protein" (Oda, 2000b, s/p). Note that informing whether a product contains the "protein of an egg" does not imply informing whether the product is GM. To the extent that nutritional composition is seen here as the main purpose of labeling, what would be informed to the consumer, in this case, is not how it was produced (process/GM product), but only its nutritional components. In the example given by the former president of CTNBio, the main concern of the labeling should not be, then, to inform the level of transgenicity of the product, but to inform its nutritional basis. In this case, once again, the view of labeling is reduced to examining the nutritional equivalence of foods. A type of information that is offered by the conventional labeling system. Which makes the requirement for GMO-specific labeling, then, unnecessary.

6.4 The Ideological Conflict About Labeling

The moral argument presented by the precautionary labeling alliance leads us to the idea of autonomy. Autonomy is related to a person's ability to make his or her own decisions based on the values that make up his or her way of life. Autonomy, then, does not refer only to the ability to make choices, but to the ability to make choices that are in harmony with the values that guide a given way of life. In this sense, an informed consumer will have more autonomy not because of the amount of information that is accessible to him, but because of the possibility of harmonizing this information with the values that correspond to his or her way of life. Thus, and as Rubel and Streiffer (2005) point out, even though many people may consider GM foods to be harmless, they may still want to avoid these products because they want to defend organic food production and the way of life that may be associated with this type of production.

This discourse not only assumes that consumers can make choices based on risk and see those choices as a right, but it also assumes that those choices could occur for reasons other than safety. For the groups defending labeling in Brazil, safety was therefore only one of the reasons to defend this policy, but not the only one. In the civil public action filed by IDEC against the release of RR soybean, it informs that the labeling is justified by factors that transcend a mere safety issue, and then it will observe that this "data [animal or plant species gene] is essential for the consumer to exercise his right to choose, considering, including allergenic, religious, cultural aspects" (IDEC, 200-? apud Prudente, 2000, s/p). This cultural vision will be evident in the dispute over the scientific or ideological character of labeling. To the extent that labeling has been associated with consumer autonomy, its absence has been seen as an attempt at ideological manipulation by those who wish to eliminate it.

In the precautionary discourse, the defense of choice and autonomy is associated with a vision of consumer empowerment. A vision that is founded on the view of the consumer as a Homo economicus who, through market mechanisms, can make well-informed choices. In this perspective, the consumer is seen as sovereign and having a dominant role in how society's resources are allocated. For the case examined here, this view of economic man suggests only the consumer's ability to use his or her power of choice as a means of influencing investments and production processes. This was the meaning given by economist Ludwig H. Edler von Mises when he stated that "the consumers are the masters, to whose whims the entrepreneurs and capitalists must adjust their investments and methods of production" (Mises,

1944 apud Klintman, 2002, p. 73). The importance accorded to labeling, therefore, is not only associated with the defense of consumer choice, but also in the influence that this choice may imply for the production and marketing of GM products. Labeling is seen as a means by which consumers can make their choices, but also because of this very possibility, of rejecting these products. In doing so, consumers could subsequently influence the economic process more generally.

Non-labeling has been seen as a threat to consumer autonomy because, in its absence, consumers may hold confusing beliefs about what they consume. In 2003 in the United States, although 70% of processed foods contained GMO ingredients, 58% of consumers believed that they had never consumed GM foods. In 2004, 41% of consumers were unsure whether genetically modified foods were accessible in supermarkets, with 11% believing they were not. Also during this period, 46% of consumers were unsure whether they had consumed GM products and 23% believed that they had not. In another US survey, only 33% of consumers knew that labeling was not required for GM products and 28% mistakenly believed that there was a mandatory labeling requirement for GM products.[1]

Part of this confusion in consumer information lies in the beliefs they hold about government regulation. In the absence of information on product labeling, consumers may believe that they are not consuming GM products. The lack of labeling may cause consumers to conclude, therefore, that in the absence of labeling information, the products are not GM products. This concern is expressed in the Brazilian case in the words of a deputy when he says that: "transgenic products from Argentina are being consumed here indirectly, and without us knowing what we are consuming". This fact would point to "a disrespect to the consumer, who should at least have the information that these are transgenic products" (Grandão, 1999, p. 36). Fernando Gabeira, federal deputy in the period, will also note that, since Brazil has been importing GM foods, it would be "necessary that these foods had an indication for consumers," because this process suggests that "Brazil continues to consume various genetically modified foods without people realizing it" (Gabeira, 2001, p. 4). These observations suggest that people would be consuming GM products not because they are in favor of them, but rather because they are unaware that they are consuming them. They would not be consuming GM products because they consider them safe, but simply because they are unaware of what they are actually consuming.

In the liberation discourse, advocating for the liberation of RR soybean, in which the defense of conventional labeling prevails, consumers have been denied a choice between these types of products, at least when specific labeling for GM foods has been disqualified. After all, to argue against mandatory, specific labeling of GM foods is to argue against consumers being able to make any kind of choice among these products. As Jacobs (1991, p. 43) will report, it is necessary to offer

---

[1] An examination of this data can be found in Streiffer and Rubel (2007).

consumers a choice that allows them to express their environmental concerns.[2] On the other hand, this marginal role of the consumer is not always inconsistent with free market ideology. This may occur since this passivity, as Klintman (2002) points out, can be accepted with the argument that consumer sovereignty is only valid in circumstances in which they, the consumers, "govern" by means of valid information. In the Brazilian case, the arguments against specific labeling because of consumers' "fears" and "mistrust" follow this direction. Therefore, the discourse of liberation involved, as will be examined shortly thereafter, a kind of political paternalism that removes from the consumer the responsibility for decision-making. In this vision, the process of rational choice is delegated to the entities and organizations that are considered to be more capable of making decisions on these aspects. In this case, CTNBio itself.

In the conventional labeling alliance, the consumer right has generally been interpreted as a right to information involving food safety. This safety is reduced in purely nutritional terms. What allowed for a consensus within this alliance is the assumption that, in terms of safety, there is no difference between conventional and RR soybean. This view was then supported by the view that sees the communication provided by labeling as a strictly scientific type of communication. It also assumed that this same scientific information was limited to nutritional safety issues. Coincidentally, these assumptions reflected the various assumptions that guided the US policy for labeling GMOs. So let us look briefly at some aspects of US policy in this area and then return to the Brazilian case.

6.5 The Policy of Substantial Equivalence

The FDA policy in the United States holds that label information is useful to the consumer only to the extent that it carries information about the nutritional basis of the food and its implications for the consumer's health. As Pariza (2007, p. 7) points out: "This position does not recognize a consumer's 'right to know' simply for the sake 'knowing,' nor does it recognize a manufacturer's 'right to inform' simply for the sake of 'informing.'" The FDA policy requires information when, and only when, that information is important to issues involving product and consumer safety. Safety is understood here in terms of nutritional safety. But to the extent that, for the FDA, GM foods are neither more nor less dangerous than conventional

---

[2] As Jacobs (1991, p. 43) points out, although orthodox economists like Mises hold that consumers are sovereign, in reality, the latter are subject to what producers offer them.

foods—at least in nutritional terms the FDA does not require specific labeling for GM foods in the United States (Streiffer & Rubel, 2007).[3]

The FDA policy reflects a change in labeling policy that has occurred in industrialized countries in recent decades. The FDA policy is guided by a Sound Science approach that reduces the information to be communicated on labels to scientific information related to nutritional safety issues. From a space intended for product advertising, labeling has been transformed in the United States into a means of providing scientific information associated with food safety to consumers. In the process, the functions of regulatory agencies have changed. More than ensuring the content of the products, the agencies began to monitor the veracity of the information contained in the labels. Cultural issues linked to the process by which products are made have been excluded from this labeling system (Guthman, 2003). From this perspective, it is not for labels to express economic, cultural, or any other kind of information that is not strictly limited to food safety issues. Since GM products are perceived as safe, at least in nutritional terms, specific labeling of GM foods is seen as providing no useful information to the consumer. Thus, the labeling is seen as involving an arbitrary separation of substantially equivalent products which, for this type of policy, is misplaced.

The discourse on the release of RR soybean in Brazil reproduced US regulatory policy lines at several points. For example, when Bresser-Pereira states that the government's policy would follow European policy, he also states that in "the case of soybeans, […] there is no substantial change of the product – the product is exactly the same […], the grain is exactly the same, indistinguishable" (Bresser-Pereira, 1999b, p. 6). In this passage, the former minister is already putting himself in line with American policy in some way by indicating that the labeling should be guided by the precept that the "grain is exactly the same." By stating that "there is no substantial change in the product" and that the "product is exactly the same." Bresser-Pereira reproduces the Principle of substantial equivalence (PSE) premise that underlies the conventional labeling policy in the United States. Former CTNBio president Esper Cavalheiro's view on labeling will also reflect the premise of the US standard that labeling should focus on the nutritional components of foods. The labeling, he will say, "of any product must provide accurate and correct information about the nutritional and compositional characteristics in order to ensure free consumer choice" (Cavalheiro, 2001, s/p). This shows that the Ministry of Science and Technology and CTNBio started to defend labeling from a precept that is precisely used to not label GM products in places like the United States where a single system

---

[3] In 2000, when asked why it did not label GM products, the FDA in the United States then made the following statement: "We are not aware of any information that foods developed through genetic engineering differ as a class in quality, safety, or any other attribute from foods developed through conventional means. That's why there has been no requirement to add a special label saying that they are bioengineered. Companies are free to include in the labeling of a bioengineered product any statement as long as the labeling is truthful and not misleading. Obviously, a label that implies that a food is better than another because it was, or was not, bioengineered, would be misleading" (FDA, 2000 apud Degnan, 2007, p. 27).

6.5 The Policy of Substantial Equivalence

is used to label GM and non-GM products from a unique labeling system. This is because the information associated with the nutritional traits does not provide any information about GMOs on food labels.

The presence of the Principle of substantial equivalence (PSE) assumptions can also be seen in the following words of the technical director of the Brazilian Association of Food Industries (Abia), who in 2005 will state that:

> GM foods have the same composition and characteristics as foods that have been consumed for years by millions of people around the world. According to the World Health Organization – WHO, in a report entitled 20 questions on genetically modified foods, such foods currently available on the market have undergone rigorous evaluations. They found that they are as safe for consumption as the others. Therefore, ABIA reiterates its commitment to transparency, credibility and respect of its members with the Brazilian consumer. (Bick, 2005, s/p)

The emphasis of the argument that conventional and GM foods have the "same composition and characteristics" is always made in the context of discussions about the labeling of GM foods. And these same claims are made while reiterating a commitment to consumer rights. And the subliminal message that is sent by the message tends to be the same. Given the existence of this similarity, the requirement for a specific labeling system for GM foods tends to be unnecessary.

The WHO was one of the first organizations to use the concept of substantial equivalence to guide its GMO labeling policy. The previous passage also indicates that consumer law has been interpreted from the conventional labeling system for GMOs. The reference to equal composition and characteristics links to the nutritional components that the Principle of substantial equivalence (PSE), advocated by the WHO, merely examines.

Moreover, in the period of the commercial release of RR soybean, Bresser-Pereira (1999b) will also report that "the American position was the same as Embrapa's." This statement can be seen as an acknowledgment that some government agencies were already aligned with the US position, more than the European one, which opts for conventional labeling (where there is no indication that the product contains GMOs) of GM foods. And in the absence of any court decision in the period, it is possible to assume that they held to this view. This position, in turn, contrasts with the information passed on by Bresser-Pereira in the same period when he stated that: "the Minister of Justice and I took to the president of the Republic the position that we should have European-type labeling, which is the one where you indicate, whenever necessary, that the product contains genetically modified product" (Bresser-Pereira, 1999b, p. 6). European policy on GM products may have been very different from American policy in many respects at the time, but with regard to labeling it was similar, or why not say substantially equivalent. It was assumed that once producers and various regulatory authorities concluded that a product was safe, then they were safe, which meant that they would be subject to a conventional labeling policy. The European position was therefore similar to the American position until 1999, when a moratorium will be declared on GMOs in the region, at which point the different European directives in this area will be reviewed. Europe will only take separate regulatory measures on GM foods after the moratorium that lasted from 1998 to 2004. In the period when Bresser Pereira makes his statement, Europe was following a similar standard of safety evaluation of GMOs as the United States.[4]

Bresser-Pereira's statement indicating that labeling policy in Brazil would follow the lines of policies in Europe and the United States is not necessarily contradictory. This is because, until the RR soybean release period, labeling policies in both places did not differ substantially. However, with the European moratorium on GMOs, the Europeans started to follow a different model of labeling for these products. This rupture occurred precisely in the year RR soybean was released in the country. What allows us to say that the labeling policy in Brazil started to express more the characteristics of the American labeling policy than the European one. Thus, if it was valid to say that until 1999 Europe and the United States had similar policies for GMOs, this statement was no longer valid after that period.

6.6 Defending the Consumer from Himself

In Brazil, the case for conventional labeling of RR soybean was justified on the basis of consumer fears and misgivings. It was seen as a way to correct the mistakes and misconceptions that consumers themselves might make in their consumption decisions. This is a characteristically paternalistic view. It is assumed that by not including this type of information, one would be benefiting the consumer himself or herself. It would be a way to defend the consumer from his or her cognitive limitations in circumstances where his or her choices would be misled.

Paternalism is not something easily definable. Some definitions see it as a restriction on an agent's freedom that is done for the agent's benefit. Other definitions emphasize the reasons why the intervention is made. Gerald Dworkin defines paternalism as "interference with a person's liberty of action justified by reasons referring exclusively to the welfare, good, happiness, needs, interests, or values of the person being coerced" (Dworkin, 1972 apud Gert & Culver, 1976, p. 45). Thus, paternalism can be understood as interference with an agent's freedom of action on behalf of his or her interest or well-being. In the American case, the defense of specific non-labeling is, for example, on a paternalistic basis. This paternalism can be expressed as follows. The public has delegated labeling decisions to the FDA because (1) the public elected congress; (2) that created the FDA (which came from

---

[4] Whiteside (2006, p. 24) notes that: "Before 1997, EU regulations - like those in the United States - pertained only to the premarket testing of GMOs. It was assumed that once GMO producers and various regulatory authorities concluded products were safe, then they were safe, period. The moratorium that began in 1999 stemmed from several countries' insistence that the EU adopt a mandatory labeling policy. Labeling requirements are precautionary in the sense that they take more seriously the possibility of regulatory lapses and unexpected developments. As noted earlier, labeling seeds and consumer products as GM is an essential condition for tracing problems, should they occur, back to their source. (…)" (2006, p. 24).

6.6 Defending the Consumer from Himself

a legitimate authority); (3) therefore, the conventional labeling advocated by the FDA should be seen as legitimate labeling; and (4) therefore, the public would be consenting to the policy implemented by the FDA.[5]

This paternalistic view emerges when it is suggested that CTNBio's decisions should be obeyed, giving it a legitimate power in its decision:

> The law, by establishing the regulatory framework, fully complies with the precautionary principle established in the biodiversity convention. The absence of scientific certainty cannot delay the application of standards, of rules. By automatically establishing this system, this regulatory logic, by creating a high-level technical and scientific collegiate body to decide whether or not there is a risk [CTNBIO], precautionary precepts are being met. (José Silvino, 2004 apud Cesarino, 2006, p. 103)

And, therefore, this same technician will say, "if there is no risk, it [RR soybean] will be treated as the common ones and will go to the inspection agencies that originally have common competences for the common ones" (José Silvino, 2004 apud Cesarino, 2006, p. 103). This argument synthesizes the vision of the equivalence discourse: for equivalent products, equivalent labeling. Note also that, in this case, conventional labeling presumes a situation in which "the precautionary precepts are being met." This paternalism is also found in Bresser-Pereira's view when he says, as noted in the previous part, that "the National Congress approved the Biosafety Law and this law established the National Biosafety Technical Commission (CTNBio)" (Bresser-Pereira, 1999a, p. 2). Therefore, also in his view, it is up to the "CTNBio to verify, case by case, whether or not a given product is likely to be approved for health and the environment, from the point of view of biosafety. [...]." This is precisely the case, for Bresser-Pereira, in the process involving the "approval and regulation of the commercial use of 'round up ready' transgenic soybeans" (Bresser-Pereira, 1999a, p. 2). For the former minister, this product "was analyzed at length and finally approved by CTNBio. Which means, in his view, that the National Congress' policy regarding transgenic products is being strictly followed" (Bresser-Pereira, 1999a, p. 2).

The paternalism existing in FDA policy in the United States tends to be reproduced here, with CTNBio as the expression of the highest authority. Replicating the example offered by Streiffer and Rubel (2004) for the Brazilian case, the following scenario is suggested: (a) Brazilians elected congress; (b) that created CTNBio, which is the legitimate authority for policy-making; and (c) therefore, the type of labeling advocated by CTNBio should be seen as having the necessary authority and legitimacy. The irony of the Brazilian case is that CTNBio offered no detailed plan for GMO labeling in 1999, while the FDA in the United States at least offered justifications for its labeling policy for products containing GMOs. In Brazil, on the other hand, the guidance in 1999 was unclear and imprecise.

The idea that CTNBio is backed by the law and that its authority should therefore be fully respected does not allow us to understand, however, the differences between

---

[5] For an examination of paternalism in the American case of policy to GM foods, see Streiffer and Rubel (2004).

the knowledge that guided CTNBio's decision and that of the critics who rejected it. This paternalism is expressed whenever the actors in favor of the release try to convince their interlocutors of their false beliefs about the safety of GMOs. In this case, what matters most is not whether or not the public has delegated all decisions to CTNBio, but whether this consent is based on the public's own knowledge. Or, as Streiffer and Rubel (2004, p. 237) point out for the American case, paternalism assumes that the public would delegate decision-making power to the FDA if "it was well informed," because it is assumed that "if people were well informed, they would change their preference by giving hypothetical consent to delegation." The words of the former president of CTNBio can be taken as an example. Some years after the release process was initiated and when asked about the relationship of trust between the public and regulatory agencies, Walter Colli, president of CTNBio in a debate sponsored by Fapesp on May 10, 2008, answered the questioning involving labeling and the relationship of trust between scientists and laymen as follows:

> When you eat organic, you are eating Bacillus and you are eating this gene, the same thing! So the only thing I want is for you to understand what a scientist is saying, that's all. Whether you are for or against is the same thing as being a Corinthian or a Palmeirense, I am a Corinthian. What am I going to do? (Colli, 2008, s/p)

The CTNBio president's answer did not boil down to this, but this was the thrust of his argument: GM foods and organic products are equal or equivalent. It is surprising, then, how representatives of the government, CTNBio, industry, and political sectors seek to demonstrate the equivalence of GM and conventional products, since this equivalence is far from the principle that governs labeling policies of GM foods in several countries, since these same policies presume the existence of the non-equivalence of these products. The public's distrust of scientists is seen as being the result of an information deficit in which the central interest, then, is to make the other understand "what the scientist is talking about." In this case, the goal is to make the public understand that GM foods are no different from conventional products and that, for this very reason, there is no reason for fuss. It is then assumed that if people were better informed, they would delegate a consent to CTNBio's decisions. People refuse GM foods not because they have good reasons to do so, but the opposite. They lack the necessary information to consider CTNBio's decisions as correct and legitimate. However, this view disregards the existing criticisms in the scientific field itself about these issues and the laymen's appropriation of this kind of critical knowledge in order to sustain their resistance.

6.7 An Environmental Utopia

One final point should be examined at the end of this chapter: the way in which systems of traceability and segregation have been incorporated into the debate on the release of RR soybean. Smith and Phillips (2002) draw a distinction between identity preservation systems, segregation, and traceability that helps us understand the

Brazilian conflict in some important respects. As these authors indicate, segregation can be seen as a regulatory tool that is needed in situations where novel foods entering the food system can create serious health risks (Smith & Phillips 2002, p. 31). Segregation will arise where, as a food safety measure, there is a concern for the mixture of the segregated product with respect to all other products. The traceability system leads us to the same question. Smith and Phillips (2002, p. 31) note that traceability systems are used when certain products present certain risks, possibly produced by bacteria or residues, need to be quickly removed from retail shelves. The traceability system allows retailers and the distribution system to identify the source of contamination and thus initiate procedures to remedy the situation. Thus, the EU sees "traceability as providing a "safety net" should any unforeseen adverse effects be established" (EU, 20--? apud Smith & Phillips, 2002, p. 32). Thus, one can consider that these systems are implemented in situations where a framework of uncertainty about the safety of goods that are produced and sold in the economic system is recognized. They are used especially in processes involving innovations in which the consequences associated with the use and consumption of what is produced involve a certain level of risk that makes the use of these systems necessary.

This makes us understand why the concerns with segregation and traceability systems have been absent for sectors of the government, CTNBio, and the food industry. These systems are nothing more than the result of a food safety policy for certain products that present a level of risk that is not very well known. And they are implemented as a food safety policy because of this very fact. So, if RR soybean is seen as having a safety level equivalent to conventional soybean, for the conventional labeling alliance, throughout the conflict, what would be the justification for implementing such systems in Brazil? To insist on such systems would be incoherent when it is claimed that the GM product is totally safe and when it is maintained that labeling is merely a political issue and not one of food and environmental safety. It is no coincidence, then, that to date the government, CTNBio, and industry sectors have not offered any detailed program for the segregation and traceability of GM products. The creation of food safety systems with these characteristics has been a banner of environmental groups and specifically of organizations such as IDEC and Greenpeace. And its implementation would require the recognition of RR soybean as distinct from and not equivalent to conventional soybean.

6.8 Final Considerations

The conflict over labeling RR soybean provides us with yet another example of how interpretive frames tend to develop in the labeling conflict. And, similar to what was seen in the previous chapters, it is possible to see patterns in the way the issue is addressed. In the case of labeling, it is possible to notice once again that the proponents of liberation offer a minimal line of rupture with the regulatory process for GM foods. One could say that, for this case, this break is even nonexistent if one considers that the labeling system advocated for GM foods for this group did not differ at all from the conventional labeling system. This view was confronted with that offered by critics of the release of RR soybean and which was especially advocated by civil society organizations. In this view, labeling was associated with environmental safety and with values that transcend mere consumer rights. It was also associated, as noted above, with the precautionary principle and the creation of a food safety system involving segregation and traceability. Thus, it is possible to see in the conflict, once again, a tension between a vision that seems to change very little the regulatory status quo and another that tends to propose more significant ruptures in the policy for GMOs.

This axis of conflict also indicates how disputes operating in different areas have reciprocal implications that are not always recognized. Thus, while CTNBio representatives argued that the labeling issue was a "much more political than technical issue," this generally overlooked that CTNBio's own views on the risks of RR soybean were repeatedly used to disqualify the need for labeling of GM products in the country. Scientific insight often carries political implications in the regulatory process, and political decisions can similarly influence how knowledge is appropriated in the process. As indicated, the political debate over labeling has at no time been dissociated from the scientific controversies surrounding its commercial release. To a large extent, this seems to stem from the very measures used to assess the risk of RR soybean. When considering traditional and RR soybean as similar or equivalent, the policy prescription tends to project naturally, taking the following form: for equivalent products, equivalent labeling.

What was seen here is not only a regulatory pattern that tends to disregard the possible economic and distributional impacts produced by the new technologies in agriculture, but also the tendency to neglect the ethical and political issues that also invade the consumption process. At the same time, the most important thing to be indicated in this debate is the democratic void expressed by the decision-making process. The debates in the CTNBio and in the government have not even come close to the discussions that take place in other parts of the world on the subject. Possibly a consequence of the technical responsibility that CTNBio attributes to itself, considering labeling a political issue that would be foreign to its attributions. On the other hand, it should be noted that many issues debated on RR soybean labeling by civil society organizations have developed out of recognition of ethical issues neglected by the government. If, on the government's side, the RA ended up justifying the defense of a conventional labeling standard, for critics, the recognition of its limitations offered support for the defense of a specific labeling system, which, in recent years, has come to prevail in the policy for GMOs in the country.

References

ANBio. (2008). Relatório sobre a liberação comercial da soja transgênica no Brasil. Disponível em: http://www.anbio.org.br/notícias/relatório.htm. 2002. Acesso em: 23 jul. 2008.

References

Bick, Léo F. (2005). Audiência pública de no. 2036/05 realizada pela comissão de meio ambiente, desenvolvimento sustentável sobre a comercialização em super- mercados brasileiros de produtos sem rotulagem. Realizada em 8 dez. 2005.

BRASIL. (2007). Projeto de decreto legislativo no. 90, de 2007. Susta a aplicação do artigo 3° do Decreto no. 4.680, de 24 de abril de 2003, que regulamenta o direito à informação, assegurado pela Lei no. 8.78 de 11 de setembro de 1990. Diário do senado federal. Maio de 2007.

Bresser-Pereira, L. C. (1999a). *Alimentos transgênicos e biossegurança. maio 1999.* Disponível em: http://www.bresserpereira.org.br/view.asp?cod=596. Acesso em: 19 ago. 2020.

Bresser-Pereira, L. C. (1999b). *Audiência pública realizada pela comissão de defesa do consumidor, meio ambiente e minorias.* No. 0533/99. Debate sobre a autorização para produção e consumo de alimentos transgênicos no país. Disponível em: https://www2.camara.leg.br/atividade-legislativa/comissoes/ comissoes-permanentes/cdc/documentos/notas-taquigraficas/not.1999.html/ nt16061999.pdf. Acesso em: 19 ago. 2020.

Cavalheiro, E. (2001). Transgênicos: sociedade precisa de informação para decidir. *Jornal da ANBio, RJ,* ano 1, n. 4, set.

Cesarino, L. M. C. N. (2006). *Acendendo as luzes da ciência para iluminar as luzes do progresso.* Dissertação (Mestrado em Antropologia Social) – Programa de Pós-Graduação em Antropologia Social. Universidade de Brasília, Brasília.

Cezar, F. G. (2003). *Previsões sobre tecnologias: pressupostos epistemológicos na análise de risco da soja transgênica.* Dissertação (Mestrado em Filosofia) – Departamento de Filosofia, Universidade de Brasília, Brasília.

Colli, W. (2008). Transgênicos e mídia. Transcrição do debate entre Walter Colli e Herton Escobar. Realizado em 10/05/2008. Disponível em: http://www.revista.pesquisa.fapesp.br/pdf/revolução_genomica/colli.pdf. Acesso em: 4 abr. 2008.

Degnan, F. H. (2007). Biotechnology and the food label. In P. Weirich (Ed.), *Labeling genetically modified food.* Oxford University Press.

Gabeira, F. (1999). Audiência pública de no. 0533/99 realizada pela comissão de defesa do consumidor, meio ambiente e minorias sobre a autorização para produção e consumo de alimentos trasngênicos. Realizada em 16 jun. 1999.

Gabeira, F. (2001). Audiência pública no. 000565/01. Comissão especial – PL No. 2905/97 – Alimentos geneticamente modificados. Realizado em 19 jun. 2001.

Gert, B., & Culver, C. M. (1976). Paternalistic behavior. *Philosophy and Public Affairs, 6*(1), 45–57.

Grandão, J. (1999). Audiência pública de no. 0533/99 realizada pela comissão de defesa do consumidor, meio ambiente e minorias. Debate sobre a autoriza- ção para produção e consumo de alimentos transgênicos no país. Realizada em 16.06.1999.

Guthman, J. (2003). Eating risk. The politics of labeling genetically engineered foods. In R. Schurman et al. (Eds.), *Engineering trouble. Biotechnology and its discontents.* University of California Press.

IDEC. (2005). *Transgênicos.* Disponível em: http://www.idec.org.br.br/files/relatório_transgênicos.doc. Acesso em: 13 nov. 2008.

IDEC. (2008a). Alimentos transgênicos: a posição do Idec. 2007. Disponível em: https://terradedireitos.org.br/noticias/noticias/idec-alimentos-transgenicos-a-posicao%2D%2Ddo-idec/. Acesso em: 19 ago. 2020.

IDEC. (2008b). Equívocos e omissões de Veja sobre os transgênicos. 28 out. 2003. http://www.idec.org.br/emacao.asp?id=499. Acesso em: 22 ago. 2008.

IDEC. (2008c). Idec contesta declarações da Abia. Disponível em: http://www.biodiversidadla.org/Noticias/Brasil_transgenicos_Idec_contesta_declaracoes_da_Abia. Acesso em: 19 ago. 2020.

Jacobs, M. (1991). *Green economy.* Pluto Press.

Klintman, M. (2002). The genetically modified (GM) food labeling controversy: Ideological and epistemic crossovers. *Social Studies of Science, 32*(1), 71–91.

Moura, C. (2001). Audiência pública de no. 000565/01 realizada pela comissão especial – PL no. 2905/97, Alimentos geneticamente modificados. Audiência realizada para esclarecer sobre o projeto de lei sob análise da comissão especial. Realizada em 19/06/2001.

Oda, L. M. (2000a). *Presidenta da CTNBio fala sobre a rotulagem dos transgênicos*. Disponível em https://www.agrisustentavel.com/trans/dialoda.htm. Acesso em: 12 nov. 2008.

Oda, L. M. (2000b). Para presidente da CTNBio, Leila Oda, é inevitável aentrada de transgênicos e a rotulagem dos produtos é a garantiado consumidor. *Matéria publicada no jornal Valor Econômico*, 5 de junho de 2000. Disponível em: http://www.agrisustentável.com/trans/dialoda.htm. Acesso em 10/4/2008.

Oda, L. M. (2000c). *Presidenta da CTNBio fala sobre a rotulagem dos transgênicos*. Disponível em http://www.radiobras.gov.br/ct/20000/materia_220900_3.htm. Acesso em 12/11/08.

Pariza, M. W. (2007). A scientific perspective on labeling genetically modified food. In P. Weirich (Ed.), *Labeling genetically modified food*. Oxford University Press.

Pizzato, L. (2000). Audiência pública de no. 0927/00 realizada pela comissão de defesa do consumidor, meio ambiente e minorias. Discussão sobre rotulagem de produtos trasngênicos. Realizada em 13/09/2000.

Prudente, A. S. (2000). Transgênicos: sentença em ação civil pública. Sentença de 26/06/2000. Disponível em: http://jus2.uol.com.br/pecas/texto.asp?id=337. Acesso em: 11 fev. 2009.

Rubel, A., & Streiffer, R. (2005). Respecting the autonomy of European and American consumer: Defending positive labels on GM foods. *Journal of Agricultural and Environmental Ethics, 18*(1), 75–84.

Silver, L. (1999). Debate. In: F. de A. P. Soares (org.). *Anais do seminário. A sociedade frente à biotecnologia e os produtos transgênicos*. Confea.

Silver, L. (2001). Audiência pública realizada pela comissão de defesa do consumidor, meio ambiente e minorias. No. 001019/01. Realizada em 27/09/01.

Smith, S., & Phillips, P. W. B. (2002). Product differentiation alternatives: Identity preservation, segregation and traceability. *AgBioforum, 5*(2), 30–42.

Streiffer, R., & Rubel, A. (2004). Democratic principles and mandatory labeling of genetically engineered food. *Public Affairs Quarterly, 18*(3), 223–248.

Streiffer, R., & Rubel, A. (2007). Genetically engineered animals and the ethics of food labeling. In P. Weirich (Ed.), *Labeling genetically modified food*. Oxford University Press.

Whiteside, K. H. (2006). *Precautionary politics. Principle and practice in confronting environmental risk*. The MIT Press.

Chapter 7
Regulation Made in the United States: Regulatory Polarization and the Brazilian Case

In the period of the release of RR soybean in Brazil, government leaders showed signs of following what was happening in Europe and the United States on the issue. In some aspects, the speech of some government representatives pointed to an intention to follow the European model of regulation. This was the case of labeling where the Minister of Science and Technology had stated that, for this issue, the country would follow this path. But this statement was made before the European moratorium on GMOs was announced. This means that Europe served as an inspiration at a time when its policy on GMOs differed little from US policy in this area. Moreover, and somewhat paradoxically, Europe also inspired those who sought to resist the release of RR soybean in this country. After all, the campaign for an "Um Território Livre de Transgênicos" sought to achieve the same economic and political effects achieved by the European moratorium. But for this case too, the closeness tends to exist more in form than in content, since this movement presented a certain radicalism that proved nonexistent on the other side of the Atlantic. But considering this scenario and taking into perspective the analysis made in the previous chapters, one can ask: What direction has the Brazilian regulatory model taken when considering the political possibilities that exist today? Would the Brazilian model be closer to the European regulatory policy? Or would it be closer to a more permissive model, such as the American case? In the remaining part of this chapter, an attempt was made to answer these questions. The question is more than valid, since the release of RR soybean, the country has not, unlike Europe, undergone any significant reformulation of its regulatory guidelines (except for the incorporation of a labeling system for GM foods). By examining these latter questions, it will be possible to clarify some of the main features of the Brazilian regulatory process for GM foods.

7.1 Regulatory Polarization

In his book *Genes, Trade and Regulation*, Bernauer notes that until the mid-1980s existing agbiotech policies in Europe and the United States were quite similar, but then differed in the 1990s. The difference reached its peak with Europe's declaration of a moratorium on GM foods. While the United States embraced agbiotech with open arms, making no significant differentiation between GM and conventional foods, regulatory policy in Europe moved to a more rigid and differentiated regulatory standard, including measures associated with the precautionary principle. The European model recognizes a clear distinction between GM and conventional foods, establishing a distinct regulatory treatment for the former. The difference between Europe and the United States is now so significant that both regions are taken to represent opposing regulatory models for GMOs. For Bernauer (2003, p. 11), the differences between the United States and Europe present the main characteristics of regulatory polarization in the world today. The regulatory polarization is expressed through two models of regulation of GMOs that take shape in the United States and Europe. What is the reason for this difference in regulatory policies in the two regions? The explanation is not simple. In Bernauer's (2003) case, the understanding of this difference lies in a set of interrelated factors. Among them are factors such as consumer perceptions, public opinion, civil society (NGO) actions, biotech company interests, farmer views, processors' and retailers' strategies, and also the functioning of the respective political systems. Thus, the controversies associated with GM foods have produced distinct regulatory models. This polarization may have important effects on the future of biotechnology. This is because polarization, as Bernauer (2003) points out, increases commercial risks for all those who participate in the global biotechnology market. It increases potential public and consumer resistance by making the acceptability of the technology less certain and predictable. These and other conditions associated with regulatory polarization in turn create a legitimacy trap for agbiotech. The industry's inability to deliver on its promises, an effect that may be produced by the polarization itself, may itself become a source of risk for agbiotech in the medium to long term.

7.2 The European Case

Unlike in other countries, and the United States in particular, the issue of GM foods in the EU presents a level of complexity that will not be found in other cases. After all, one is not making reference to a country, but about a political association that brings together the interests of several nation-states. This makes the European case unique in many respects, because unlike elsewhere, the conflict over the release of GM foods has developed against a rather complex political backdrop. And this backdrop is something that must be considered when examining regulatory policy for GM foods in Europe. The moratorium expresses, in a way, an effort to create a common policy among countries with specific political, cultural, and economic characteristics. To understand the existing regulatory policy on GM foods in Europe, it is necessary to understand the development of this policy in two phases. The first phase makes up Europe's regulatory policy in the period leading up to the moratorium on GM foods that will be declared in 1999. The second phase emerges with the moratorium itself and the policies that have developed since.

In the 1980s, *agbiotech* was positively integrated into Europe's future plans. In this vision, the European region should join the trends that structured the new knowledge society in a context of economic globalization[1]. It became evident to the European Commission the need for a safe and competitive integration of the region in an environment of globalized markets. To this end, Europe perceived the need to strengthen its internal economic capacities, but also the need to remove existing trade barriers with those countries (United States) that could become important economic partners for the region. In this context, it was also important to remove existing regulatory obstacles within Europe itself in order to create an attractive environment for investments in strategic economic and technological sectors. Finally, environmental concerns were allied to this vision. The new policy guidelines should operate in tune with the imperative of sustainability that, since the 1980s, was strengthened within the European public opinion and in international forums organized by the UN in the name of sustainable development.

Agbiotech was seen as an important piece in this vision that Europe established for itself. First, it was seen as an important component in building a knowledge society for the region. Therefore, investment in agbiotech was seen as part of this project. Thus, agbiotech, representing a major technological trend of today, was seen as a vital area for Europe's global economic integration. Policies that were supposed to remove regulatory barriers for sectors considered strategic usually made reference to investment barriers within agbiotech itself as well. The regulatory standardization that the rapprochement between Europe and the United States required should cover trade involving agbiotech itself. A regulatory standardization between these countries should then be promoted in order to enable trade associated with the products generated by the economic incorporation of agbiotech in agricultural systems.

As Levidow (2014a) indicates, this vision of agbiotech did not occur only in discourse, but was expressed in specific actions. One of the practical effects of this vision was the gradual subordination of research in science and technology to the priorities of the private sector, as a way to attract investment to areas considered strategic, as was the case of agbiotech. This vision was also articulated in the existing negotiations between the United States and Europe to generate harmonization in the regulatory standard to be adopted by both regions for this area. This was also expressed, for example, in the European approach to issues associated with the patenting of living organisms. Finally, research funding has encouraged closer relationships between companies and universities, creating an equally favorable climate for agbiotech in the region[2]. It will therefore not be uncommon for some papers examining European regulatory policy for agbiotech in this period to see the features of

---

[1] Our considerations of the European view of agbiotech in the pre-moratorium period is based largely on the works of Levidow (2014a, b) and Levidow and Carr (2010). A detail of the issues that will be examined here can be found in different papers published by the author on this period.

[2] However, it seems that some internal tensions between EU member states began to emerge in this period, since, as Levidow (2014a) points out, some of them began to disapprove of the policy guidance offered by the European Commission. And it seems that the decision to release GM seeds without significant changes in the regulatory process will end up intensifying these conflicts that have remained latent among EU member states.

a neoliberal policy. For it was as lax as that in the United States. As in the American case, it was guided by a narrow risk assessment and was at the same time convergent with the interests of agbiotech promoters, without establishing any regulatory orientation that resembles what is required today on European territory. Levidow (2014b) summarizes some of the characteristics of European regulatory policy in this period with the following words:

> Thus early EU regulatory procedures incorporated policy assumptions of the agbiotech promoters. Under "risk-based regulation," societal decisions on agbiotech were reduced to a case-by-case approval of GM products, within a narrow definition of risks, placing the burden of evidence mainly upon the objectors. (Levidow, 2014b, p. 835)

Beyond this economic vision associated with agbiotech, the European policy vision for agbiotech was perceived as converging with environmental concerns. This component of the European Commission's discourse was very present in the 1990s in the run-up to the moratorium in 1999. Agbiotech was perceived as an opportunity to reconcile agbiotech with the ecological imperative. According to Levidow, the EU policy by appropriating the language of sustainable agriculture to address GMOs "attributed agri-environmental problems to genetic deficiencies, while attributing eco-efficiency benefits to the inherent properties of GM crops" (2014a, p. 156). In this period the European Commission endorsed the vision of biotech companies that suggested the possibility of a win-win game between economics and ecology. Agbiotech was thus seen as a type of clean technology that would create new economic opportunities. Therefore, agbiotech would allow combining economic efficiency with the demands of regional environmental sustainability. One of the main reasons reinforcing this view was the belief that agbiotech would help promote resource use efficiency. For, after all, among some of the benefits that are perceived to be associated with GM foods, as seen in Chap. 3, are reduced use of water resources, less use of pesticides, higher agricultural productivity, and less pressure on forest areas.

However, this does not mean that there was no concern for safety issues associated with Europe's commercial use of GMOs. In a sense, as Levidow (2014a) points out, the European regulatory framework developed out of a tension between its liberal economic impetus and the safety issues associated with agbiotech. According to him, this framework was oriented in an attempt to integrate concerns associated with product safety with a project to overcome trade barriers between member states (Levidow, 2014a, p. 157). Thus, the EU will apparently guide its regulatory policy by increasingly foregrounding safety issues, although it left open what it considered safe and dangerous in the realm of innovations from agbiotech (Levidow, 2014a). However, as the European Commission began to incorporate growing concerns about safety issues and to demonstrate a willingness to adopt a more preventive policy, the institutions responsible for the governance of GMOs in the region became the focus of strong pressure from interest groups representing agbiotech companies. The European regulatory framework came under attack from multinationals hostile to attempts to incorporate stricter safety measures. These interest groups considered these measures a threat to European competitiveness due to the

7.2 The European Case

regulatory burden they could produce. According to Levidow (2014a), this criticism then led to an ebb in biosafety concerns, causing the European Commission's plans to be scaled back in such a way that a more permissive approach to regulating GMOs ended up being imposed. This approach merely emphasized the need for RA, leaving aside issues associated with the PP.

The first sign of the most intense conflict that would ensue in Europe came with the European Commission's approval of a variety of GM corn. The case made evident the disagreements that existed between member states over the regulatory policy for GMOs that was being implemented (Lee, 2008, p. 02). And because of the controversy, some governments began to draft for themselves an alternative regulatory approach to GMOs. Approaches that sought to transcend the adopted regulatory guidelines. Then, in the second half of the 1990s, the image that many Europeans held of GM foods was reversed despite the discourse linking agbiotech to a type of ecological modernization. GM foods came to be seen as associated with the narrow interests of large corporations and the perception of their risks began to outweigh the debate about their potential benefits. Popular campaigns emerged that were critical of the commercial planting and use of GM foods on European territory. And these campaigns took over the Internet and the television media, galvanizing the support of public opinion. Even before the moratorium was declared, several events took place in 1998 and 1999 that signaled public dissatisfaction with the existing regulatory policy that came into effect in Europe. Plantations with GM seeds were destroyed and were followed by the mainstream media. Similar protests took place in France, where they sought to communicate to the public the uncertainties and dangers associated with GMOs.

Until the declaration of the moratorium in the late 1990s, Europe had been going through a slow and gradual process in its attempt to devise a regulatory framework for GMOs. However, in a relatively short period of time, this entire framework was perceived as inadequate (Lee, 2008, p. 02). Then, in 2001, the first outlines and principles of the new European regulation for GMOs emerged with the *Deliberative Release Directive*. With this directive came the first laws applicable to all GMOs. In 2003, a second piece of legislation also emerged with the *Food and Feed Regulation*. This period saw the emergence of rules applicable more directly to GM foods and also to traceability and labeling issues. In examining the consequences of the change produced by the moratorium in Europe, Lee (2010) notes that:

> Protection of human health and the environment are enormously important and very complex, but so may be the way agricultural biotechnology distributes risk, benefit, and power, globally and locally, and the profound uncertainties about its physical and socio-economic impacts. The new regulation of GMOs allows for the importance of capturing factors beyond risk, apparently accepting that the authorization of GMOs is not simply a technical decision. There are opportunities for public participation, and for the consultation of an ethical committee; there is even reference (somewhat oblique) to the socio-economic impacts of GMOs. And in respect of food and feed GMOs, there is an explicit recognition that "scientific risk assessment alone' may not provide all the necessary information, and that a decision can be based on 'other legitimate factors'." (Lee, 2010, p. 104)

With the new regulatory framework, Europe has moved away from the principles that are generally associated with the US regulatory model. Among the characteristics of the current regulation of GM foods in Europe, the following can be highlighted: (a) guarantee of property rights, (b) specific law for GM foods, (c) emphasis on the production process of GM foods (not only on the product), (d) separation of responsibilities in risk management (analysis, communication, and risk management), (e) broad environmental risk assessment, (f) discrimination of GM foods, (g) application of the precautionary principle, (h) specific and mandatory labeling of GM foods, (i) traceability system, and (j) coexistence rules.

With the change, the European regulatory process now operates as follows[3]. Any and all types of GMOs produced by agri-biotech, whether food or feed, require authorization from the EFSA (European Food Safety Authority) to be commercially released[4]. Second, the applicant for authorization must notify the responsible member state for the authority to examine the application and provide an institutional response. In its response, the competent authority must prepare a report presenting its own assessment, which must be shared with other member states. The report produced by the competent authority must offer a response, where it will give its verdict on whether the product in question may be launched on the market. Thus, when the applicant for a GM innovation sends its application to the national competent authority, a risk communication process starts, for which EFSA is responsible. EFSA receives the application and then passes it on to the European Commission, which then shares the information with the member states.

An important aspect of this process concerns risk assessment. Where regulatory decisions were previously based solely on a technical RA, and this in turn was seen as sufficient to guide EU decisions on the release of GM foods, the new European policy orientations have brought significant changes to this process. Risk assessment tends to be more complex in its new format. After the moratorium, the *Food Law* Regulation established a division between risk analysis, risk management, and risk communication. Risk assessment is then broken down into three different processes. In the case of risk assessment involving GM foods, EFSA needs to consult all national competent authorities (Lee, 2010, p. 107). If applications for the use of GM products occur, EFSA should establish communication with the national authority requesting and receiving an environmental risk assessment from the national authority (Lee, 2010, p. 107). As the author notes, Regulation (EC) 178/2002 will place EFSA as "an independent scientific point of reference in risk assessment" and its main responsibility lies in carrying out risk assessment, with "risk management in principle for political institutions" (Lee, 2010, p. 107). Thus, although EFSA centralizes in some aspects the risk assessment, this process tends to involve the member states, giving conditions, in this process, for the

---

[3] The following details of the regulatory process are based on the work of Lee (2008, 2010).

[4] Although there are different standards and guidelines for the two cases (food and feed), there are requirements that are common to both. For example, both products require an environmental risk assessment and must follow the same procedures to carry out this assessment. For more details on this process, see Lee (2008, 2010).

"incorporation of national perspectives on risk assessment" (2010, p. 107). At the same time, the policy decision expresses an autonomy from the data and conclusions that can be drawn in the risk assessment, hence, apparently, the separation between risk assessment and risk management. The process is more complex and dialogic and, in some cases, may involve negotiation between the parties involved. Especially with regard to studies and risk analysis of GMOs.

7.3 The American Case

Controversies associated with agbiotech also emerged in the United States in the 1990s. The first concerns arose with the use of the growth hormone for dairy cattle known as bovine somatropin (bST). BST became the first GM product used for food production in the United States and, as Priest (2005) recalls, was the first political test of agbiotech in the country. In the case of bST, criticism of the hormone's use came from small farmers in states like Vermont and Wisconsin. The use of bST created fear on the part of these farmers of possible economic loss. Criticism also came from environmentalists who feared the chemical effects of the hormone on animal welfare on farms (Priest, 2005). In the 1990s, other controversies were associated with the commercial release of Bt cotton, Bt corn, and GM StarLink corn. The first two are genetically manipulated by receiving the genes from the bacterium *Bacillus thuringiensis*. With genetic manipulation, GM corn starts to express insecticidal proteins for certain types of insects. The genetic transfer of the genes from this bacterium into Bt cotton and Bt corn created the expectation that plantings of these two varieties of plants would require less insecticide use. However, critics argued that the use of these seeds could lead to a loss of crop effectiveness over time as insect resistance develops through natural selection. Concerns by some farmers arose over the fear that use of these GM plants could then end up compromising organic farming itself (Priest, 2005, p. 35)[5].

Despite these controversies in the recent history of the United States, the existing literature is somewhat consensual in pointing to the favorable institutional environment that the country ended up creating for the commercialization of GM foods. Currently, the United States is one of the global leaders in the development and production of GM seeds. It is the region that has created the most permits for the testing, planting, and commercial use of GM seeds (Bognar & Skogstad, 2014, p. 73). The number of companies that exist in the country exceeds the number that

---

[5] In the case of StarLink GM corn, the discussions took place for other reasons. Environmental groups made a complaint indicating that traces of transgenic corn, which had not been released for human consumption, had appeared in products from the Taco Bell chain. The complaint prompted the FDA and EPA authorities to open an investigation into the case. In the case of the United States, the political and economic controversies over the use of GM foods were not restricted to the national sphere, but also expressed in international politics. In 2002, the United States provided food aid to Africa, but this ended up involving an international conflict over the viability and reception of the food, since it was denounced for being GM food (Lee, 2008, p. 20).

can be found elsewhere. US companies account for 2/5 of the world's patents for GM products. This has had the effect of changing the profile of what is produced in the country. By 2010 the country comprised almost half of the world's production of GM crops (Bognar & Skogstad, 2014, p. 71). By 2020, GM soybean accounted for 94% of all soybean grown in the country, and GM cotton and GM maize accounted for 96% and 93%, respectively[6].

There are several factors that help explain this US dominance in GM food production. All of them are associated with cultural factors pertaining to the US political reality itself. First, a pro-biotechnology coalition has emerged in the United States involving industry, agriculture, the food industry, and politics. A coalition that gained direct access to policy decisions in this area. The political group, which was sensitive to the demands of this coalition of actors, was made up of elected politicians who were particularly sensitive to the interests of large farmers (Bognar & Skogstad, 2014, p. 76). Second, this context was promoted by the Republican governments of the period, which, with neoliberal policies, provided the necessary support to build a permissive regulatory framework for GMOs. Third, the pro-business culture in the United States reinforced the tendency to adopt a more permissive regulatory framework for GMOs. Fourth, and unlike elsewhere, regulatory policy for GM foods has gained support from a substantial proportion of US farmers. Unlike in Europe and elsewhere, for example, where farmers were more divided. Finally, beyond this internal context, one could add the international alliances that ended up reinforcing the pro-GMO policy in the United States[7].

Responsibility for the regulatory process in the United States is divided among three agencies: FDA, EPA, and USDA. The first is responsible for safety and labeling. The EPA regulates GM seeds that are associated with the use of pesticides. The USDA and its department of agriculture (APHIS) take responsibility for safety from field trials. These agencies should work in harmony, but there is no shortage of work suggesting that this is not always the case. Within this division of roles between these agencies, GMO developers must also provide information on the safety of GM foods. They must provide information indicating the absence of risk if their

---

[6] The updated figures for GM food production in the United States can be found on the US FDA website. See, for example, https://www.fda.gov/food/consumers/agricultural-biotechnology

[7] The United States and Canada have exerted a strong influence in international organizations like the OECD in the context of trade policies associated with GMOs. As they have converging interests in agricultural policy, the United States and Canada have sought to shape international laws in a way that favors their commercial interests. Also during this period, the United States was the major proponent of TRIPS, the agreement that protects the interests of biotechnology companies (Bognar & Skogstad, 2014, p. 73). The "blue book" produced by the OECD established three main axes for GMO regulatory policy. All of these still hold true today in US GM food policy. Among these guidelines established by the commission are the principles that (a) in terms of risk there is no substantial difference between GM products and conventional products, (b) genetic manipulation techniques are considered to express safety in obtaining results, and (c) in scientific terms, the construction of specific legislation for the use and marketing of GM products is seen as of little importance. For a brief analysis of the OECD guidelines for GMOs, see Lunel (1995).

products are commercially released. And companies bear a substantial share of the risks associated with these products.

As Bognar and Skogstad (2014) point out, regulation in the United States tends to be permissive. It supports and satisfies the interests of agbiotech companies without creating many bureaucratic difficulties for them. The process offers guarantees for the ownership of GMOs and security for commercial release. In this way, the projection of the United States is associated with its permissive regulatory framework that does not discriminate between GM products and normal products. The main axes that characterize US regulatory policy emerged in the 1980s, specifically in 1986, when the country joined the OECD's GMO policies. In 1992, the FDA declared its intention to treat GM products in the same way as products containing additives that are protected by traditional methods (Bognar & Skogstad, 2014, p. 74). Although the agency stated that GM products should reach the same safety standard as other foods, it signaled in building regulatory process for GMOs where no special rules would be created for them (Markert & Backer, 2003, p. 49). GM products would be evaluated by the same safety standards as conventional products. Thus, the United States declared that most GM products were presumptively GRAS (generally recognized as safe).

7.4 Brazil, Europe, and United States

In the case of Brazil, when the regulation of GMOs is examined in formal terms, based on its laws, it would be possible to consider the country as having a strict regulatory framework for GMOs. This could be the view of those who consider that the country has adopted reasonably strict controls for the regulation of GMOs through the adoption of measures such as labeling. However, this is a type of assessment that emerges unaware of the regulatory conflicts examined in this book. And also unaware of how labeling came to be implemented in the country. If the labeling of GM foods is disregarded, many similarities between the United States and Brazil can be seen. The similarity between the two cases is striking in many points. Almost all the elements that are associated with US policy can be found today in the regulation of GM foods in Brazil.

But before taking some of these convergences into account, let us consider the context through which the two regulatory frameworks emerged, as it may shed light on some issues in some ways. As considered above, the US regulatory framework developed from a set of factors that helped create its permissive regulatory policy for GMO foods. It was the result of the influence of a political coalition that favored pro-market regulation of GM foods in the country. In the Brazilian case, the context tends to be closer to the US case for similar reasons. In Brazil, the regulatory structure will emerge under the Cardoso government, which took a liberal stance on the economy. And it is possible to find in his government the same objectives of creating a pro-market context for agbiotech innovations as occurred in the United States and Europe in the pre-moratorium period. The Cultivar Law, which helped fuel the conflict in the south of the country, is the result of an economic policy that sought to create an environment of legal security in the country aimed at investment in areas such as agbiotech. A context that the leftist governments that followed did nothing to change. Thus, biosecurity laws in the country emerged in parallel with economic policies of this type that approach agbiotech in strictly economic terms[8].

At the same time, in addition to this factor, one should consider in this context the influence of the political coalitions that exist in the United States and Brazil, which are similar in some respects. As seen above, Bognar and Skogstad (2014) note that the regulatory framework in the United States was strengthened by a broad coalition that included biotechnology developers, farmers, food processors, members of the public sector (state and federal), and even scientists from public universities. The authors thus signal that the political pressure exerted by this coalition was successful thanks to its members' privileged access to the political decision-making process and their direct contact with policy makers. In this last group are political leaders who, due to the financial support received in their electoral campaigns, were sensitive to the interests of the large landowners. In Brazil, a similar reading can be found not only in research addressing the release of RR soybean in the period, but it also emerges in the views of scientists who served on the CTNBio itself. Thus, Leonardo Melgarejo, a former member of this committee, in an interview in 2013, evaluated the country's GM food policy from the following perspective:

> It is a bet of transnationals, conveyed through agribusiness connections, not the government itself. The change of governments, in this field, has not brought any differences. FHC, Lula and Dilma allowed and allow those interests to enforce their objectives. In other words, in my view, the government ends up being guided by agribusiness, which defines its strategic option and makes it viable through its agents, who operate inside and outside the government. Whether or not there is a political-ideological option of the current government for this model, the significant presence of ruralists in Congress reinforces a game of give and take that interests the predominant model of agriculture. A small group of companies owns the technologies, their patents, and the distribution channels for seeds, pesticides, and agricultural machinery and equipment. These companies act together and their strength prevents the government from making independent decisions on matters that concern them. (Melgarejo, 2022)

Next, the former CTNBio member observes that: "In fact, what occurs is that in this field government options seem contaminated by the options of agribusiness, which in turn responds to the interests of large transnational corporations" (Melgarejo, 2022).

While this question has not been a central focus of our analysis in this book, it needs to be inserted here in order to put into perspective the context through which the regulatory structures that exist today in Brazil and the United States have

---

[8] As Paarlberg (2001) observes, intellectual property right protection policies emerged in the 1990s incorporating a permissive rather than preventive approach to GM seeds. This more permissive environment then encouraged the interest of biotechnology companies to invest in the country. Then, as Paarlberg (2001, p. 69) writes: "Strengthened IPR guarantees have also helped encourage research on GM crop technologies by Brazil's own agricultural scientist working inside state-funded institutes. If Brazil fails in the end to participate in the GM crop revolution, weak IPR policies will not have been the reason."

7.4 Brazil, Europe, and United States

developed. If the former member of CTNBio is correct, then the basis for understanding some of the elements of regulatory policy in Brazil should begin with an analysis of the political economy of agbiotech. After all, it would be naive to imagine that what is considered today the "flagship" of the Brazilian economy would act in a manner uninterested in decisions about agbiotech in the country, if one considers, especially, its current influence on the Brazilian political process. Some aspects of this process became evident in our analysis, as they are also evident in other studies on the regulation of GMOs in Brazil. In the US case, one of the factors that favored its regulatory framework was the fact that farmers embraced GM seeds as part of their economic strategy. In the Brazilian case, the same convergence can be found. Large farmers, linked to agribusiness, have embraced GM seeds as a business strategy, unlike small farmers who, in many cases, are affiliated with alternative agricultural strategies. This was evidenced in the conflict in the south of the country, where the campaign for an "Um Território Livre de Transgênicos" was supported by farmers investing in a model of family farming and agroecological farming. Large farmers, in turn, took up the defense of RR soybean.

Putting these contextual factors aside and looking at the elements that characterize regulatory policy for GM foods more specifically, the Brazilian and American models can now be examined. Some of the essential elements that constitute the American model of regulation can be found in the Brazilian case. In order to examine these issues, let us return once more to some of the characteristics of the American model. It was possible to see in the previous part that some of the characteristics that can be associated with this model are the following. There is (a) strong property rights protection there for GM products. At the same time, there is no (b) discrimination against GM products in US policy. This makes it possible to give these products the same regulatory treatment as is offered to conventional products. In the United States, therefore, products from genetic engineering (c) are not considered to create any significant differential risk when compared to conventional products. In addition, the (c) authorization of GM products is guided by a RA that focuses on environmental and human health aspects (nutritional characteristics). Regulation of these products also operates through the (d) principle of substantial equivalence. This provides support so that both products (GM and conventional) receive the same treatment through the regulatory process. This process is accompanied by (e) industry self-regulation. Self-regulation is considered a reliable instrument to promote the safety of GM products, as companies assume for themselves the risks associated with the innovations brought about by agbiotech. Which explains the state's lack of commitment to taking more rigid preventive measures in this area. The RA incorporated into the regulatory process embraces only issues associated with (f) health and the environment, which means, as Bognar and Skogstad (2014, p. 74) note, that "ethical and socio-economic considerations are not considered in the formal regulatory regime."

When considering the analysis of the release of RR soybean in Brazil, it will be possible to find many of the elements found in the US regulatory model. In the process of commercial release of RR soybean, RR soybean was approved through a RA that sought to assess the nutritional safety of the product. A procedure that is also adopted in the United States. The release of RR soybean was also based on the application of the Principle of substantial equivalence (PSE), another strong orientation of the existing regulatory process in the United States. And these measures, as seen above, contrast with the precautionary approach existing in Europe, as these criteria are considered insufficient to validate the release of GMOs on European territory.

When one considers the speeches of the members of CTNBio in the process of releasing RR soybean, one will find the same elements that exist in the regulation of GM foods in the United States. CTNBio members have repeatedly suggested that GM foods present no differential risk compared to conventional foods and have generally used this argument to justify the release of RR soybean as it occurred, as it was possible to see with respect to the evaluation of a technician that justified CTNBio's decision in the period "if there is no risk, it [RR soybean] will be treated as the common ones and go to the inspection bodies that originally have common competencies for the common ones" (José Silvino, 2004 *apud* Cesarino, 2006, p. 103). These arguments were also used to delegitimize the need of a specific labeling for transgenics in the country. At least it is clear that, throughout the conflict, RR soybean was considered equivalent to conventional soybean. Hence, it is difficult to clearly demarcate discrimination and precaution as elements of Brazilian regulatory policy. For if the precautionary principle can be reduced to the application of RA, then not only would Brazil's regulatory policy reflect a precautionary approach, but the other countries that restrict their safety measures on the basis of this criterion. In that case, there would be a few cases, if any, that could be classified differently. Thus, the way in which RR soybean has been released reflects the US rather than the European model.

Bognar and Skogstad (2014) state that the RA used in the United States covers only environmental and health issues. No criteria other than these factors can or should be considered when assessing the risks of GM products. In the previous section, where the European case was considered, it was possible to see how the changes produced by the moratorium in Europe led the region to reconsider this issue. In the words of Lee (2010), with the new regulatory policy in Europe, there is now "an explicit recognition that 'scientific risk assessment alone' may not provide all the necessary information, and that a decision can be based on 'other legitimate factors'" (Lee, 2010, p. 104). Therefore, taking these differences between the American and European models in perspective, what conclusions can be drawn in relation to the Brazilian case on this issue? Once again, a significant similarity between the American and Brazilian models can be seen here. The RA that is conventionally applied by CTNBio in Brazil follows the same precepts as the American policy. As seen in the case of RR soybean, distribution issues associated with small farmers, or any other risks that go beyond environmental and nutritional safety issues, were disregarded by CTNBio. The committee's considerations incorporated nothing beyond what conventional RA might encompass, a procedure that has been maintained in all existing releases in the country. This procedure, in turn, as already seen, is part of the American regulatory model.

Regulation of GMOs: United States, Europe, and Brazil

Policy regulatory	United States	Europe	Brazil
Property rights	Yes	Yes	Yes
Product/process	Product	Process	Product
Specific law for GM foods	No	Yes	Yes
Discrimination	No. Policy does not differentiate between GM and conventional products.	Yes. Policy establishes a difference between GM foods and conventional products.	(?) Ambivalence in existing policy. Principle of equivalence sees GM and conventional foods as equivalent. Labeling policy, in turn, acknowledges the difference.
Principle of substantial equivalence	Yes	No	Yes
Risk (scope)	Health and global environment	Health, environment, and other "legitimate factors"	Health and environment
Precautionary principle	No	Yes	(?) Policy relies on application of Principle of substantial equivalence (PSE) and RA. Contradictions existing in regulatory policy. For the application of precaution is not seen as being reducible to these principles.
Analysis, communication, and risk management	Concentration of functions at the FDA. Responsible for risk analysis, communication, and management.	Separation of functions. EFSA is responsible for risk analysis and risk communication. European Commission remains responsible for the management (policy) of the risk.	Concentration of functions in CTNBio. Responsible for risk analysis, communication, and management.
Labelling	No Products are subject to the conventional labeling system.	Yes Indicative of the existence of genetic manipulation. Labeling associated with coexistence policies and product segregation.	Yes Indication of the existence of genetic manipulation. Labeling dissociated from a policy of coexistence and segregation of products.

Policy regulatory	United States	Europe	Brazil
Labeling and precaution	No. Conventional labeling disconnected from the precautionary principle.	Yes. Specific labeling and linked to risk management. New regulatory framework makes post-market environmental monitoring possible.	No. Differential labeling, not linked to the risk management of GM food. Lack of post-market monitoring.
Coexistence	No	Yes	No
Traceability	No	Yes Traceability covers all stages of commercialization.	? Conventional traceability system. There is no specific traceability law for GM foods.

Source: Author

The US model does not provide labeling for GM products, which is not the case in Brazil. Thus, there would be a significant difference between the United States and Brazil in the regulation of GM products. Labeling would be proof that precaution is being applied. The difference does exist, but it should be put into perspective with the analysis made in this book. It should be noted that the labeling currently in place in Brazil is not the result of actions taken directly by CTNBio. Nor has it been applied with a view to environmental safety considerations (or associated with the precautionary principle), although it may partially accomplish this purpose. A question that will soon be addressed below. So much so that CTNBio itself did not draft any line favorable to labeling during the RR soybean release period, since it considered that safety conditions were being satisfactorily respected with the completion of the RA Labeling did not emerge as a result of application of the precautionary principle, but as application of the consumer's right to choose what he or she wants. Hence, labeling should be considered permissive, since it did not emerge as part of a risk management policy on the part of CTNBio. Franke et al., in examining the existing labeling policy in Brazil and the EU, note that:

> No specific laws exist on the traceability of GMOs and derived products in Brazil. In the EU on the other hand, traceability is obligatory in general according to the General Food Law (EC Regulation No. 178/2002). Each member state of the EU is obliged to check their food and feed supply for compliance with EU food and feed law, including GMO legislation (Control Regulation EC/882/2004 and amendments). (Franke et al., 2009, p. 34)

In turn, when examining the Brazilian case, the authors note that whereas the labeling policy in the EU is "process-based," meaning that all GMOs, or products derived from GMOs, need to be labeled, "In Brazil, the labeling policy is 'product-based', which implies that only if recombinant DNA or transgenic proteins are present in the end-product, it is subject to labeling regulations" (Franke et al., 2009, p. 34), indicating that the labeling in Brazil presents significantly different characteristics from those present in the European policy. In Europe, labeling is perceived as an

7.4 Brazil, Europe, and United States

extension of risk management to GM foods and as a way of applying the precautionary principle. European labeling allows GM products to be monitored and inspected after they are placed on the market. This in turn makes it possible for these products to be part of a more extended risk management, since these studies can be made possible after the commercial license has been granted. To this end, European labeling presumes a policy of coexistence and segregation of GM seeds. Brazil's labeling policy, on the other hand, does not embrace these issues. And it is far from improving the labeling system so as to transform it into an integral element of risk management as the European model follows. In June 1999, under pressure to implement a labeling policy, the government responded by announcing a commission to address the issue. In examining this process, Paarlberg (2001) shows that the initial proposal only incorporated "unpackaged fresh foods currently sold without labels or food additives, food preparations, or processed foods" (2001, p. 86). Paarlberg (2001) also notes that the government document was written carefully so as not to require costly steps to segregate GM from non-GM foods in the Brazilian domestic market. On the other hand, civil society organizations began to demand a precautionary labeling policy that suggested the need for segregation (Paarlberg, 2001, p. 86). In the end, the government's more permissive approach prevailed. The more precautionary labeling policy proposed by organizations like IDEC ended up not prevailing. An attempt to give the Brazilian labeling system a precautionary bias occurred again in 2005 through Bill 4809/05, sponsored by Congressman Edson Duarte of the Green Party (PV). IDEC, in publicizing the bill at the time, describes it as follows:

> Bill 4809/05, presented by Federal Deputy Edson Duarte (PV-BA), provides for the monitoring of the effects of genetically modified foods and products, even when already released for consumption. The effects of these products on the environment, the human organism and animals must be analysed. According to the proposal, the public inspection bodies will be responsible for establishing specific monitoring plans for each product (...). (IDEC, 2022)

The bill, however, was not passed into law. It should be noted that, if such a system were created in Brazil, it would require a reconfiguration of the regulatory bodies, since, as IDEC reports, these bodies would need to set up monitoring plans for GM foods. This is apparently what the labeling policy in Europe assumes. It is quite possible that the proposal was based on the new GMO policy guidelines that exist there, as the proposal features the various elements that are present in the European model. Consider that the MP's bill will emerge after the Europeans have made public their new policy for GM foods. What marks Brazilian politics in recent years is not the tendency for the country to move in this direction, but the opposite. What seems to be gaining increasing strength and support from dominant political groups is the effort to try to totally eliminate specific and mandatory labeling for GM foods in the country. Hence, labeling has not become an enduring achievement and is currently a target of those who seek to nullify its effects[9]. This may occur in the coming years.

---

[9] On recent attempts by the agribusiness caucus to nullify the effects of mandatory labeling of GMOs in Brazil, see Cortese et al. (2021).

The permanence of the labeling policy for GM foods in the country today is highly uncertain.

There is a difference between GM food labeling systems that operate on the basis of the usual strategies for promoting food safety, where the aim is to examine only the nutritional safety of foods and those that aim to broaden, as is the case in Europe, the very notion of safety from a new risk management perspective. In Regulation (EC) No. 1830/2003, the European Commission states that, besides enabling the possible removal of products that have adverse effects, the traceability system should also: "facilitate the implementation of risk management measures in accordance with the precautionary principle." As Grossmann (2007, p. 44) notes, Regulation 1830/2003 establishes then a link between labeling, traceability, and risk management. This document establishes a unified traceability system for GM products which is seen as aiming at "facilitating accurate labeling, monitoring the effects on the environment and, where appropriate risk management measures including, if necessary, withdrawal of products" (Regulation EC 1830/2003, 2003, p. 25). Similarly, Aarts, Rie and Kok (2002, p. 69) point out that the new EU regulation, besides offering the consumer's "right to know," also facilitates "enforcement of regulatory requirements and are of importance for environmental monitoring and post-market surveillance"[10].

So while the labeling system in Europe may operate as a support for differentiated risk management, this is not possible in Brazil, for this would require changes in the direction that the bill of deputy Edson Duarte of the Green Party (PV) envisaged. This difference between one model and another does not occur, then, only because the labeling systems present different characteristics in terms of information, but also because these systems operate by different orientations within the respective risk management policies. Labeling, in the Brazilian case, was not seen as part of the risk management system at any time by its main regulatory body (CTNBio), at least, nothing beyond what the usual labeling systems can offer, so much so that the labeling emerged over the heads of CTNBio members and not as part of its risk policy for GM foods in the country. Therefore, the two labeling systems are embedded in distinct regulatory guidelines in the context of the risk management of GM foods. The labeling of GM foods in Brazil emerged as a result of pressure from civil society rather than policies deliberately implemented by CTNBio. As Paarlberg (2001) puts it:

---

[10] One must consider that a traceability system is, by definition, a type of policy associated with risk management. So, as Ene (2013) notes, the "main purpose of traceability development is to increase security and safety throughout the food chain and to establish an acceptable model for raw material supply (...)" (Ene, 2013, p. 289). And one of its main objectives is to "manage risks related to food safety and animal health issues" (Ene, 2013, p. 289). However, a more detailed analysis of the relationship between labeling, traceability, and risk management in European policy can be found in Devos et al. (2012). In it, the authors also examine the post-market environmental monitoring plans (PMEM) that are proposed by the European Commission in its new regulatory guidelines for GM foods.

7.4 Brazil, Europe, and United States

> Public demand for GM food labeling in Brazil was a direct outgrowth of the 1998 lawsuit against the commercial release of RR soybeans. As noted above, it was the Brazilian consumer advocacy organization IDEC that initiated the lawsuit, and one of IDEC's key arguments was Brazil's lack of any GM-specific food labeling policy. All subsequent court rulings on the case consistently mentioned labeling an essential condition for lifting the court-ordered suspension. (Paarlberg, 2001, p. 85)

This seems to indicate that all the elements that bring the Brazilian regulatory model closer to the US model emerge from the direct decisions of the CTNBio and groups that work on behalf of agribusiness interests, while the criteria that could bring it closer to the European model result from a process of judicialization of GMO policy in the country. If it were not for the pressure exerted by civil society, the country would hardly have implemented any labeling policy for GM foods. However distant this labeling system is from the existing European policy in this area. And this would have occurred with the consent of CTNBio itself, which, throughout the process, showed itself indifferent to the issue. This shows that regulatory policy tends to operate through a conflict between the state and civil society. It is the state's representative bodies that take a permissive view, while society has sought to impose criteria more closely associated with precaution. The country's regulatory policy has emerged from this conflict. At the same time, the permissive nature of the labeling system is due to these tensions that project onto regulatory policy for GM foods.

Another element that is bringing the Brazilian model closer to the US model concerns the processes associated with risk analysis, communication, and management. As seen in the previous part, with the new GM food policy, the European Commission established a clear division between risk analysis and risk management. EFSA, the main regulatory body in Europe, is now responsible for risk analysis and risk communication. Within the scope of risk analysis, EFSA is the institution that provides scientific advice on the topic. Some important features in this dimension are the existing transparency in the process. The opinions of the commission are made public, with the exception of personal data linked to property rights. However, EFSA's scientific assessments relating to foreseeable health effects can never be kept secret (Alemanno, 2008, p. 13). Another important point is that EFSA does not have the final word on GM foods when there are profound scientific divergences. In Alemanno's (2008) words: "The Authority lacks formal authority to reach binding resolutions on potentially contentious scientific issues, it does not have the final word in case of diverging scientific opinions between its own decisions and those issued by other bodies." While EFSA has the responsibility for risk assessment and risk communication, policy decisions (management) are the responsibility of the European Commission itself. When examining this distinction between analysis and communication and, on the other hand, risk management, Alemanno (2008, p. 22) observes that by establishing such a division, the European Commission seeks to promote coherence and transparency between the functions of analysis, communication, and risk management.

If the European model, then, has led to a division of labor between these dimensions (analysis and management), the Brazilian model reiterates the American model once again. There as in Brazil, the functions are exercised by a single body.

In the American model, these functions are exercised by the FDA. And the same process tends to occur in Brazil, with CTNBio centralizing decisions related to risk analysis, communication, and management. Although there was a suggestion that these functions be separated in organizational terms at one point in the Brazilian conflict, the opposite has prevailed: the centralization of CTNBio's decision-making. CTNBio and FDA are not only characterized by their centralized decision-making on GMOs, but also by the assumptions they use to guide their policies.

The European model is currently less decentralized and more dialogical. In the case of risk assessment involving GM foods, EFSA needs to consult all national competent authorities (Lee, 2010, p. 107). At the same time, the division between risk analysis and risk management allows the decision-making process to consider factors beyond the data provided by the usual risk analyses. Even because this analysis is unable to examine economic or other issues that fall outside its scope. In the models that exist in Brazil and the United States, management decisions (policies) end up taking on this same narrowness. Therefore, with respect to the relationship between science and policy, the Brazilian model can also be seen as closer to the American model.

Finally, one last point to consider, in the United States, industry self-regulation is perceived as a reliable instrument to create the safety of GM foods. As such, self-regulation allows the government to transfer responsibility for the risks associated with GM foods to business. Which means that in the US case, companies are required to minimally assume the possible risks associated with the release of GMOs. In the Brazilian case, it is not clear that this is occurring. Which, in this case, may indicate the existence of even greater weaknesses in Brazil's GM food policy when compared to the US model. If this is true, the Brazilian regulatory model may, in theory, be operating in a framework of greater institutional irresponsibility when compared even to the American model. Since the American model tends to operate through industry self-regulation that partially transfers risk responsibilities to agbiotech companies, it is not very clear that this is happening in Brazil, which may indicate greater weaknesses in the "Brazilian way of life" of currently regulating GMOs.

7.5 Final Considerations

So, in a nutshell, what does Brazil have in terms of GM food regulation? Simplifying a little, it could be said to be the American model, with the addition of a permissive labeling system. Permissive because the labeling is not allied to any precautionary policy. Its main objective was to respect the consumer's right to choose. This does not mean that, in critical situations, this system could not operate so as to fulfil the usual role assigned to conventional labeling systems. But in this form, it provides a safety policy for GM foods that remains at the limits of the safety provided by conventional labeling schemes. This system does not allow, as already mentioned, risk management as it is implemented today in Europe. However, for the reasons given above, labeling is not part of a policy of precaution being promoted by the main regulatory body for GM foods in the country. It is a labeling that was imposed by the will of civil society through judicial means and not as a result of prior planning by CTNBio itself.

References

Aarts, H. J. M., Rie, J. P. F., van & Kok, E. J. (2002). Traceability of genetically modified organisms. *Expert Review of Molecular Diagnostics, 2*(1), 69–77. https://doi.org/10.1586/14737159.2.1.69

Alemanno, A. (2008). The European Food Safety Authority at Five. *European Food and Feed Law Review, 3*(1), 2–24.

Bernauer, T. (2003). *Genes, trade and regulation. The seeds of conflict in food biotechnology.* Princeton University Press.

Bognar, J., & Skogstad, G. (2014). Biotechnology in North America: The United States, Canada and Mexico. In S. J. Smyth, P. W. B. Phillips, & D. Castle (Eds.), *Handbook on Agriculture, Biotechnology and Development* (pp. 71–85). Edward Elgar.

Cesarino, L. M. C. N. (2006). *Acendendo as luzes da ciência para iluminar as luzes do progresso.* Dissertação (Mestrado em Antropologia Social) – Programa de Pós-Graduação em Antropologia Social. Universidade de Brasília, Brasília.

Cortese, R. D. M., et al. (2021). Reflexões sobre a proposta de modificação da regulamentação de rotulagem de alimentos transgênicos no Brasil. *Ciência & Saúde Coletiva, 26*(12), 6235–6246.

Devos, Y., Craig, W., & Schiemann, J. (2012). Transgenic crops, risk assessment and regulatory framework in the European Union. In R. A. Meyers (Ed.), *Encyclopedia of Sustainability Science and Technology* (pp. 10765–10796). Springer.

Ene, C. (2013). The relevance of traceability in the food chain. *Economics of Agriculture, 60*(2), 287–297.

Franke, A. C., Greco, F. M., Noordam, M. Y., Roza, P., Eaton, D. J. F., Bindraban, P. S., & Lotz, L. A. P. (2009). *The institutional and legal environment for GM Soy in Brazil.* Publisher Plant Research International, Wageningen University & Research Centre.

Grossman, M. R. (2007). European community legislation for traceability and labeling of genetically modified crops, food, and feed. In: P. Weirich (Ed). *Labeling genetically modified food: The philosophical and legal debate.* Oxford University Press.

IDEC. (2022). *Venceu o lobby dos transgênicos.* Publicado em abril de 2005. Disponível em https://www.idec.org.br/uploads/revistas_materias/pdfs/2005-04-ed87-politicas-biosseguranca.pdf. Acesso em 03.03.2022.

Lee, M. (2008). *EU Regulation of GMOs. Law and decision making for a new technology.* Edward Elgar Publishing Limited.

Lee, M. (2010). Multi-level governance of genetically modified organism in the European Union: Ambiguity and Hierarchy. In L. Bodiguel & M. Cardwell (Eds.), *The regulation of genetically modified organisms.* Oxford University Press.

Levidow, L. (2014a). European Union policy conflicts over agbiotech: ecological modernization perspectives and critiques. In S. J. Smyth, P. W. B. Phillips, & D. Castle (Eds.), *Handbook on agriculture, biotechnology and development* (pp. 153–165). Edward Elgar.

Levidow, L. (2014b). EU Regulatory conflicts over GM food. In D. M. Kaplan & P. B. Thompson (Eds.), *Encyclopedia of food and agricultural ethics* (pp. 834–841). Springer.

Levidow, L., & Carr, S. (2010). *GM food on trial. Testing European democracy.* Routledge.

Lunel, J. (1995). Biotechnology regulations and guidelines in Europe. *Current Opinion in Biotechnology, 6*(3), 267–272.

Markert, L. R., & Backer, P. R. (2003). *Contemporary technology. Innovations, issues, and perspectives.* The Goodheart-Willcox Company, Inc.

Melgarejo, L. (2022). *A transgenia está mudando para pior a realidade agrícola brasileira* Entrevista com Leonardo Melgarejo. Disponível em https://www.ecodebate.com.br/2013/06/04/a-transgenia-esta-mudando-para-pior-a-realidade-agricola-brasileira-entrevista-com-leonardo-melgarejo/. Acesso em: 10 de abril de 2022.

Paarlberg, R. L. (2001). *The politics of precaution. Genetically modified crops in developing countries*. The Johns Hopkins University Press.

Priest, S. H. (2005). Biotechnology. In *Science, technology, and society. An Encyclopedia* (pp. 34–40). Oxford University Press.

Chapter 8
Conclusion

The present study sought only to examine the dynamics of the conflict over the release of RR soybean in the country, which began in the late 1990s and has continued in subsequent years. Therefore, it did not attempt to provide an updated assessment of the conflicts over GMOs for the present moment. The issues examined here began at the end of 1998 and extended over several years. All the considerations made here are limited to this historical period. Therefore, the present study does not cover possible conflicts that have occurred after this period. It should be noted that GM crops have become a reality in Brazilian agriculture today and that the specific labeling system for these products has been incorporated into the country's environmental policy. A second study would be needed to examine how the policy on GMOs in agriculture develops today on national territory. This would make it possible to develop an updated assessment of the achievements and defeats that the campaigns carried out by civil society organizations have achieved throughout this process. The implementation of a labeling system for these products seems to be one of them. Additional research would also allow us to examine how the regulatory process of GMOs in the country has been developing in comparative terms with other countries in the world. Due to the limits of the present work, no policy recommendation has been offered here either to address the conflicts examined. Answering these questions would make the work run away from the narrow objectives that were set for the analysis. Having made these considerations, some of the main points that have influenced the conflict over GMOs in the country can now be considered.

The conflict that occurred in the south of the country shows us how it began by rehashing old issues associated with the green revolution. Family farming and agroecology were seen as representing an intrinsically virtuous structure of relations in environmental (sustainability) and social (justice) terms. When technological innovations are perceived as obstacles to the achievement of these ends, they come to be perceived as enemies to this kind of ethical vision associated with different agricultural modes. This indicates that, in the case of RS, GM seeds were seen as an extension of a process of rationalization ("green revolution") that threatened the virtuous practices of a new agricultural model that was being envisioned by the government in the region. By preventing the promotion of this virtuous agricultural social structure, the RR soybean came to be seen as an obstacle to the realization of an alternative agrarian reform agenda. Thus, the answer to understanding the conflict in Rio

Grande do Sul lies in the Dutra government's policy agenda for agrarian reform and how this agenda was integrated with the agrarian conflicts that evolved in the region during this period. The conflict not only reflected these tensions but also induced their radicalization, giving them, in a certain way, a new shape.

With regard to the conflicts that have unfolded over environmental risks, it is possible to see that critics of RR soybean release have not only placed the precautionary approach in tension with the RA, seeing the former as a more robust perspective for facing the scientific uncertainties existing in the decision but have also, for that very reason, established a difference between the RA and EIA. The controversy over the RA presented by Monsanto highlights the dispute over the scientific status of studies that seek to assess environmental risks linked to GMOs. In these cases, it is common for scientific knowledge to have greater authority in the debate to the detriment of other forms of knowledge. Thus, if EIA and RA are seen as producing reliable scientific data, they will also be accorded the same degree of respectability associated with the usual scientific discourse. The possible scientificity that can be attributed to the RA and EIAs in some way also confers authority for the definition of the risks associated with GMOs and the political and economic decisions that are associated with them. However, as examined, the information that EIA and RAs provide cannot be seen as providing a secure and definitive scientific basis for guiding policy decisions. This is why, for critics of these environmental policy instruments, the emphasis on each one of them appears to be merely the influence of a reductionist technocratic rationality. For these critics, the restriction of environmental policy decisions to aspects associated with EIA only tends to reveal the hegemony of this type of rationality in the environmental decision-making process. Even though it is necessary, the EIS would also be a decision-making instrument that would be far from being considered sufficient to guide the deliberation on these issues. This would mean, in this case, that linking the commercial release of GM foods to formal EIA compliance would already imply, in a way, a favorable decision for these products (Schlindwein, 2004, p. 344). If these critics are right, then by conditioning the release of RR soybean on the use of an EIA, the precautionary discourse would also be running the risk of reproducing some of the assumptions in the opposing discourse. It seems to suggest that EIA would be free of the same problems associated with the RA and that conflicts could be circumvented with the greater scientific rigor provided by these studies. The clash of science against science, as Congressman Fernando Gabeira in the period interpreted the conflict, could be captured by the image of a dog chasing its own tail.

If the release discourse represents the application of a minimal precautionary approach, this model tends to contradict the most robust views of this principle. These views first recognize scientific uncertainty as an essential condition for the application of the principle and disprove the use of the RA itself as an exclusive and unique resource for dealing with this uncertainty. The precautionary approach, as Stirling and Mayer (2000, p. 346) point out, "acknowledges the difficulties in risk assessment by granting greater benefit of the doubt to the environment and to public health than to the activities that may be held to threaten these things." So, the most usual understanding of PP puts it in conflict with RA when this type of analysis is

8 Conclusion

seen as a first and sufficient condition to provide a satisfactory response to the most complex environmental risks. But this conflict could cease to exist if the RA were seen only as an integral, rather than sufficient, part of the precautionary approach.

For advocates of a specific labeling to foods containing GMOs, it has been associated with environmental safety and values that transcend simple consumer law. It was also associated, as it was possible to see, with PP and the creation of a food safety system involving segregation and traceability of GM foods. This vision was confronted with that offered by the government, CTNBio, and the food industry. The arguments presented by these actors not only offered dubious support for the "consumer's right," often suggesting that this could be respected with a conventional labeling system, but relied on the Principle of substantial equivalence (PSE), which is rightly used elsewhere to invalidate specific labeling for GM foods. The premise of substantial equivalence underlies much of the argument rejecting labeling for GMOs. Yet the labeling of these products presumes a break with this principle, since these policies rely on the assumption that these products are not substantially equivalent. Among the countries with specific legislation for labeling, the only commonality among them is the widespread requirement that GM products be labeled because as perceived, these products, as Gruère and Rao indicate, "are not substantially equivalent to their conventional counterparts" (2007, p. 52).

In these countries, labeling is seen as mandatory because it is recognized that consumers must be informed about the new properties of products so that they can make choices in line with their own interests (Gruère & Rao, 2007, p. 52). The application of Principle of substantial equivalence (PSE) is therefore counterproductive to support a labeling policy for these products. The principle invalidates a basic premise of the labeling policy for GM foods, which states that these products are different from others and therefore deserve different treatments in their labeling. It is this differentiation that underpinned the creation of the country's biosafety law and the creation of CTNBio itself, which, ironically, came to treat RR soybean as an equivalent product to others in the period examined in this study.

Finally, in the last chapter, it was possible to verify the similarity between the regulatory policy in the United States and the one that was implemented in Brazil. The use of the RA, the application of the Principle of substantial equivalence (PSE), and other elements that were present in CTNBio's discourse and decisions reflect the US policy for GM foods. Even in the area of labeling, which would bring the Brazilian system closer to the European model, it is possible to find the elements that bring the Brazilian model closer to the American model. After all, at several moments, it was seen as unnecessary in CTNBio's view. And it only ended up being implemented due to the pressure exerted by civil society. And in light of the constant attacks that the labeling system has been suffering in the country more recently, CTNBio seems to offer little support for the maintenance of this policy in the medium and long term. And if the existing labeling systems in Brazil and Europe are compared, it is possible to see significant differences in the way these policies have been implemented in each case. In Europe, the system is seen as integrating the precautionary principle and a risk management that should be operated by the principle bodies responsible for regulating GMOs. In the Brazilian case, the system operates in terms of the more conventional labeling systems.

The Brazilian regulatory model has operated through an existing tension between a minimal regulatory vision, set in motion by an economic, scientific, and political elite that directly influences CTNBio's main decisions, as opposed to a more preventive vision that can be found in the law and in the demands of civil society and even from a minority group of scientists who are members of CTNBio and who, repeatedly in recent years, have shown themselves to be contrary to the policies implemented. The law, as seen in the case of labeling, was left at the mercy of pressure from civil society to implement it in the country. If the labeling of GM foods demonstrates a victory for civil society in the process of regulating GM foods in Brazil, on the other hand, it is indicative of the fragility of the foundations of this policy in some way. Coincidentally, the Brazilian policy is quite convergent with the economic interests of agribusiness that tends to reduce biotechnological agriculture to a merely economic issue. This also occurs in the United States. Therefore, it will not be strange to verify that the Brazilian policy has become quite similar (equivalent?) to the American policy in several dimensions, regulatory guidance that extends, by the way, also to the use of pesticides and other issues that are linked to the current agricultural policy in the country. Therefore, it does not seem strange to say that the regulatory policy for agricultural biotechnology in Brazil today is closer to the American model than to the European one.

References

Gruère, G. P., & Rao, S. R. (2007). A review of international labelling policies of genetically modified food to evaluate India's Proposed Rule. *AgBioForum, 10*(1), 51–64.

Schlindwein, S. L. (2004). Introdução de plantas transgênicas no Brasil: basta avaliar o risco ambiental? *Cadernos de Ciências e Tecnologias, 21*(2), 343–353.

Stirling, A., & Mayer, S. (2000). Precautionary approaches to the appraisal of risk: A case study of a genetically modified crop. *International Journal of Occupational and Environmental Health, 6*(4), 296–311.

Index

A
Agrarian reform, 4, 48, 49, 56, 58, 68, 71, 72, 74, 75, 78–89, 163, 164
Agribusiness, 152, 153, 158, 159, 166
Agricultural biotechnology, 25, 28, 31, 32, 61, 73, 74, 147, 150, 166
Agroecology, 49, 57, 62, 71–75

B
Biotechnology, 3, 19–36, 42, 55, 56, 63, 65, 68, 73, 85, 95, 98, 124, 144, 150, 152
Brazil, 1, 3–5, 9, 16, 25, 26, 29, 35, 38–40, 42, 43, 48, 52, 53, 55, 58–60, 68, 69, 72, 75, 76, 87, 89, 101, 104, 106, 110, 115, 119, 123, 124, 128, 129, 131, 132, 134, 136, 137, 139, 143, 151–160, 165, 166

C
Choices, 5, 29, 37, 77, 103, 112, 119, 121, 126, 128–134, 136, 165
Consumer law, 135, 165
CTNbio, 4, 16, 40–42, 55, 56, 59, 60, 62, 65, 66, 69–71, 84, 89, 93, 95, 96, 98–101, 103–105, 107, 108, 110–114, 121–127, 129, 130, 133, 134, 136–140, 152–156, 158–161, 165, 166

D
Discourses, 2–5, 7, 9–16, 21, 23, 42, 43, 48, 53, 55, 57, 62, 63, 68, 72, 74–76, 85–88, 93, 96, 98, 101–105, 107, 108, 110, 113, 114, 119, 122–128, 131–134, 137, 145–147, 164, 165
Distributive conflict, 31–32

E
Environmental discourses, 11, 14, 60
Environmental impact assessment (EIA), 32, 41, 49, 60, 64, 94, 95, 110–112, 114, 115, 130, 164
Environmental justice, 3, 16, 51–54, 56
Environmental policy process, 7–16
Ethics, 3, 19, 29–31, 36, 37, 96
Europe, 5, 28, 35, 38, 39, 49, 63, 69, 76, 96, 103, 110, 127, 135, 136, 143–148, 150–161, 165

F
Family farming, 49, 57, 58, 61, 62, 71–75, 77, 87, 153, 163
Frames, 2, 3, 9–16, 66, 93, 109, 139
Framing, 3, 9–11, 13, 94, 120

G
Genetically modified organism (GMO), 1, 3–5, 12, 16, 20, 21, 24, 30–33, 35, 36, 38–40, 42, 43, 47–49, 52–73, 75–77, 81, 89, 96–100, 102, 104–115, 120–130, 132–137, 139, 140, 143, 144, 146–151, 153–157, 159, 160, 163–166

GM foods, 4, 19, 20, 23, 25–38, 40–43, 62, 64, 65, 77, 96, 105, 107, 115, 122, 124–136, 138, 139, 143, 144, 146–148, 150–161, 164–166

L

Labeling, 3–5, 28, 29, 35, 37–38, 41, 56, 99, 115, 119–140, 143, 151, 154, 156–161, 163, 165, 166

P

Policy decisions, 36, 113–115, 127, 149, 150, 159, 164
Precaution, 5, 93–95, 98, 101, 103, 105, 106, 108, 109, 111, 112, 114, 119, 120, 128, 129, 154–156, 159, 161
Precautionary principle, 40, 94, 100, 109, 119, 120, 128, 136, 139, 144, 147, 148, 154–158, 165
Principle of substantial equivalence, 5, 94, 95, 100, 103, 120, 153–155
Public controversies, 20–23, 27

R

Regulations, 3–5, 7, 15, 19, 28, 33–38, 49, 56, 68, 94, 98, 104–106, 108, 110, 122, 123, 132, 135, 137, 143, 144, 146–149, 151–158, 160
Regulatory polarization, 35, 144
Risk analyses, 4, 35, 49, 99–102, 109–114, 120, 148, 149, 155, 156, 159, 160
Risks, 3–5, 7, 19, 21, 23, 26–30, 33, 35–37, 41, 43, 48–52, 54–56, 58, 60, 62, 63, 66–68, 70, 71, 75, 83, 93–115, 119, 120, 123, 125, 126, 128–131, 137–140, 144, 146–149, 151, 153–161, 164, 165
Roundup ready (RR) soy, 1, 39, 47, 101, 103, 129
RR soybeans, 1, 3–5, 7, 16, 23, 29, 38–42, 47–89, 93–115, 119–140, 143, 152–154, 156, 159, 163–165

S

Scientific uncertainties, 4, 93–95, 104–106, 109, 111, 164
Story lines, 3–5, 9, 11, 12, 16, 43, 119

CPSIA information can be obtained
at www.ICGtesting.com
Printed in the USA
LVHW082021080123
736723LV00006B/331